PCB Design Guide to Via and Trace Currents and Temperatures

For a complete listing of titles in the
Artech House Microwave Library
turn to the back of this book.

PCB Design Guide to Via and Trace Currents and Temperatures

Douglas Brooks
with Johannes Adam

ARTECH HOUSE

BOSTON | LONDON
artechhouse.com

Library of Congress Cataloging-in-Publication Data
A catalog record for this book is available from the U.S. Library of Congress.

British Library Cataloguing in Publication Data
A catalogue record for this book is available from the British Library.

Cover design by Andy Meaden meadencreative.com

ISBN 13: 978-1-63081-860-9

685 Canton Street
Norwood, MA 02062

10 9 8 7 6 5 4 3 2 1

Contents

Preface

Printed circuit board designers have lots of things to think about. Typically, the first is just connecting the dots. Then, how to do so on the minimum number of layers. Then, there are all the signal integrity issues to deal with. And then, finally, how do we size the few power traces there are for carrying the required current?

This latter question constitutes only a very small fraction of the total job. But it can be the most mysterious, in part because there aren't many reliable guidelines to work from. Then, many of the guidelines that exist are incomplete, contradictory, or, even worse, flat wrong. This book is intended to cover the entire topic of the relationship between PCB trace and via current/temperature relationships. It covers:

- The theories behind the various aspects of the topic;
- The results of simulations based on those theories;
- In several cases, summaries of experimental results based on, and confirming, the simulations.

To start, here are the typical questions we have received through the years, and where appropriate, where the answers will be in the text.

1. *How hot should my trace be*? The answer is, "I don't know!" How hot a trace should be is a *policy* decision. It is made by the person filling the *system engineer* role in the project. It is not made by the person filling the *board design engineer* role, even though those may be the same people. How hot a trace can or should be depends on the application. A medical product (especially if placed inside the body)

or a product designed for outer space will have a much different reliability specification than will a cable TV box designed to sit in a family room. Therefore, the thermal requirements will be much different. But somebody needs to answer that question before the board design starts.

2. *How hot will my trace be?* The temperature of a trace depends on several factors. Here are some of them and where they are discussed:

   - The current flowing down the trace. (We talk about current almost everywhere in the book!)
   - The resistance of the trace. This starts with resistivity and the trace dimensions (width × thickness). We cover this in Chapter 3.
   - The trace form factor. Wider, thinner traces will be cooler than narrower, thicker traces (Chapters 5, 6, and 7).
   - The thermal conductivity of the board dielectric. This has a *major* impact on the trace temperature because it relates to how well the board material conducts heat *away* from the trace. We cover this in Sections 2.2, 4.2, and Chapters 6 and 7.
   - Radiation and convection. Collectively these combine into something called the heat transfer coefficient (HTC). See primarily Chapter 6.
   - Other environmental factors. The existence of such things as parallel traces, underlying planes, pads, and even trace length all tend to lower the trace temperature. We look closely at these things and try to quantify them in Chapter 7.

3. *How hot will my via be?* This is the big surprise in the book. Via temperatures do *not* depend on the current through the via or on the via dimensions. Via temperatures depend primarily on the temperature of the parent trace. See Chapter 8 for a discussion of the theory behind this conclusion, via simulation results, and experimental verification of this important finding.

4. *Can I use a trace as a fuse?* In general, the answer to this is no. If a fuse is appropriate, then use a fuse component. But there are cases where this question has relevance. For example, consider a trace that normally carries a nominal current but in the case of a catastrophic system failure it may be subject to a substantial overload. It needs to be able to carry that overload for 1.0 second in order for there to be a safe shutdown of the system. How big does the trace have to be? We don't want to design the trace to carry the overload indefinitely, just for 1.0 second. Chapters 11, 12, and13 answer this question and provide guidelines. (Note: A board on

which any trace has been subjected to a very high temperature should be considered as damaged and replaced.)

5. *How does frequency or wave shape affect the relationships?* Not surprisingly, it appears that simply using the root-mean-square (RMS) value of the signal for the current magnitude is the correct approach. Chapter 16 discusses this theory, some simulations, and a few experimental evaluations in more depth. Similarly, we assume the standard resistance corrections for skin depth are also appropriate, but that is beyond the scope of the text.

6. *What is the relationship between current and current density?* Current density is not an independent variable. It is a derived variable equal to current divided by cross-sectional area. Current is the relevant variable. Current density is not. *Stop thinking in terms of current density*. Chapter 15 explains why.

7. *When can I use simple charts for the current/temperature relationships and when must I use computer simulation tools? What tools exist and what do you recommend?* Charts give you a starting point and in most cases result in worst-case (most conservative) estimates. But when you are dealing with more complicated systems and need more precision, field effect (computer-generated) solutions are required. (We have been there before with controlled impedance trace calculations.) I am not in a position to recommend any specific programs. Dr. Johannes Adam has collaborated with me since 2016 on this project and has provided me with a license for Thermal Risk Management (TRM) software. It is the only thermal simulation software I am familiar with and that is why it plays a prominent part in this book. The simulations in this book use only a small fraction of TRM's overall capabilities. Other programs exist, but I am not familiar with them.

8. If one asks the question, *"How do I start evaluating my trace requirement,"* here is one possible outline:

    - You *must* know the maximum current (and perhaps the duty cycle) the trace will have to carry. It is the circuit design engineer's responsibility to determine this. If you don't know this, nothing else matters!
    - Someone must determine the maximum allowable temperature increase of the trace. This is a policy variable; there is no right or wrong answer. Typically, it is the system design engineer who sets this policy. The policy will probably depend on the end market, the degree of reliability desired, and perhaps outside regulatory agencies.

- Given steps 1 and 2, select the trace size from existing graphs and tables. This will likely (but not necessarily) result in a worst-case size; that is, one larger than necessary.
- Determine the resistance of the trace. There are lots of calculators that can do this. You can estimate it with the following formula:

$$R = .68 \times L/(W \times Th)$$

  where

  R = Resistance in ohms at 20°C

  L = trace length in inches

  W = trace width in mils

  Th = trace thickness in mils

  Resistance will increase approximately 4.0% per 10°C rise in temperature.

- Calculate the voltage drop across the trace using ohm's law (V = current times resistance).
- You now are in position to take one of three actions:
  - Increase the trace size in order to lower the voltage drop across the trace (not a likely requirement).
  - Leave the trace size as determined.
  - Decrease the trace size because there are reasons to know this trace has enough thermal margin to do so. This is when a simulation tool is most helpful.

This final choice is what this book is about. This choice does not depend on the circuit or the power distribution network (PDN). It depends on the materials used and the environment around the trace. And the impact of the environment is a very tricky thing to get our arms around. Unfortunately, there are no cookie cutter answers—no formulas, no charts, no tables that can provide all the answers. By the time you get through this book you will understand why that is.

Technical Note: TRM

This book is written primarily for PCB designers, people who want to know "How big does my trace have to be in order to handle the current it is carrying?" Naturally, this question is asked before the board is fully designed; it is too late to ask it *after* the board has been designed!

In this book we have made liberal use of the Thermal Risk Management (TRM) software tool to analyze thermal issues of and around individual traces, and the tool is very well suited to doing this. But this may leave the reader with the mistaken believe that that is the limit of what TRM can do. In fact, TRM is *much* more powerful than this!

Dr. Johannes Adam originally conceived and designed TRM to analyze temperatures *across* an entire circuit board, taking into consideration the complete trace layout with optional joule heating as well as various components and their own contributions to heat generation. Although the program could be adapted to the measurement of an individual trace, as we have done here, it was not originally conceived with that use in mind. So when we describe how to enter the dimensions of a single trace into the software using one of the menus (see Figure 6.7), that obscures the fact that an *entire board* can be entered into the software using, for example, the entire set of Gerber and drill files, a much simpler and more accurate (not to mention more complete) process.

Throughout the book we show many images of thermal diagrams the software produces relative to a single trace under analysis. But Figure 1 illustrates how a similar thermal profile can be generated of an *entire* board, including components, under load. Imagine the power of this for a systems or package designer worrying about the thermal performance of an actual

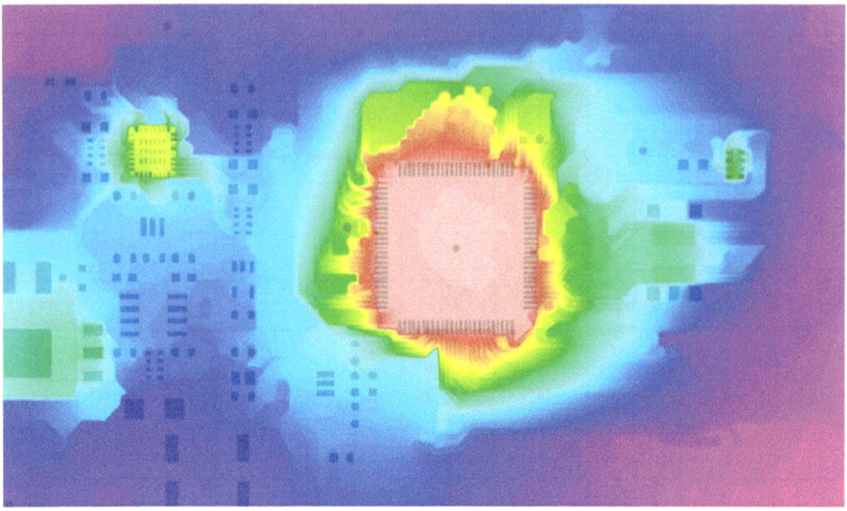

Figure T.1 TRM image of a major portion of an entire board.

product in a difficult environment (such as the engine compartment of a car, the confines of a very small enclosure, or the extremes of outer space!)

TRM is a complete, total system-wide tool for thermal analyses. We are using (and demonstrating) only a small part of its capabilities.

You can learn more about TRM and all of its capabilities at: http://www.adam-research.de/en/.

Acknowledgments

This book would not have been possible without the generous support of a lot of people. Lots of authors say this, but boy it is really true in this case.

In particular, I want to thank my longtime partner Dave Graves (now with Monsoon Solutions in Bellevue, Washington) for helping prepare the final artwork for the various test boards.

Adam Harris and Sarah Ackerman at C-Therm Technologies (Fredericton, New Brunswick) graciously measured the thermal conductivity of some board material to facilitate the via simulations. Their work is described in Appendix A.

Fabio Visentin and Alejandro Golob of Jesse Garant Metrology Center, Windsor, Ontario, provided x-rays of a board and showed me how to analyze those x-rays. Their help made Chapter 16 possible. Priority Labs (Richardson, Texas) allowed us to use Figure 16.1 as an example of how x-rays are used in failure analyses.

Scott Dau, a Seattle fireman and part-time fire investigation instructor, provided valuable assistance in focusing in on localized hot spots on traces even at very low temperatures. Norocel Codreanu, of Politehnica University of Bucharest (Romania) provided support in looking at whether there are hot spots on traces (there are) and what the thermal characteristics of right-angle corners are. Chapters 12 and 13 could not have been written without the support of these people.

Nitin Bhagwath, product architect at Mentor Graphics with the high-speed simulation group since 2012, sat in on one of our presentations and immediately started thinking outside the box. He is the inspiration behind Chapter 10.

Nathaniel Peters of W. M. Keck Center for Advanced Studies in Neural Signaling provided valuable support in microscopy.

And a very special thanks to numerous people at Prototron Circuits, Inc. (of Tucson, Arizona) who provided the test boards and also the micro-sectioning work and measurements for the various experimental tests. At the risk of accidentally leaving someone out, I want to particularly include Dave Ryder (president), Jerome Larez (Sales), Mark Thompson (Engineering Support) and Kevin Neumann (CAD, CAM, QA). Their support was fundamental in making this book possible.

But most of all, my collaborator on trace thermal issues, Johannes Adam (Leimen, Germany), was the person who started it all back in 2016. His interest, support, and encouragement (and tolerance) was what allowed me to start pursuing avenues that led to far more destinations than I ever expected. I could not have gone down this path without him.

Thank you all.

The success of this book (whatever it may be) belongs to all these people. And I hope there are no significant errors. But if there are any errors, they are the result of my imperfect understanding and in spite of everyone else's efforts.

Douglas Brooks
February 2021

CHAPTER

1

Contents

Introduction and Historical Background

1.1 Bottom Line

- The first known study of printed circuit board (PCB) current/trace relationships was a preliminary study released in 1956.
- Our industry did not get reliable data on trace current/temperature relationships until publication of IPC-2152 in 2009.
- Before 2009, the previous data kept us out of trouble because it was so conservative.

1.2 Historical Background

The first study of the relationship between trace currents and temperature is believed to have been done in 1956. It was published by the National Bureau of Standards as Report # 4283.[1] There is a good summary of this effort contained in the IPC publication IPC-2152.[2] The empirical data was not very well controlled (because of limited resources), and the resulting charts, when published, were labeled "Tentative." The authors recommended that funding be provided for a more detailed, more carefully controlled study, but such funding was never forthcoming.

Originally, there were two sets of charts, one for external traces and one for internal traces. The experimental data only applied to the external traces. The internal trace charts were derived by the researchers by simply derating the external charts by a factor of two, on the expectation that the internal traces would not cool as well as the external traces would, and would therefore be hotter.

Through the years the charts were redrawn and republished, and somewhere along the line the word "Tentative" was dropped.

Apparently they were first published as part of MIL-STD-1495[3] in 1973. They were later published in a subsequent standard, MIL-STD-275E,[4] published in 1984. Later still the charts were published as part of an IPC standard IPC-D-275,[5] and still later in IPC standard 2221.[6] Copies of some of these source documents can be viewed on the internet.[7]

These charts and standards were considered as the bible for trace temperature estimates, in spite of their humble beginnings (and the fact that the assumption regarding internal traces would turn out to be quite incorrect!) They were used by most printed circuit board designers. In hindsight, the best thing they had going for them was the test of time. They apparently were appropriately conservative because few board failures have been traced back to using them.

Finally, the IPC helped sponsor a very thorough study on trace currents and temperature that was released as IPC-2152 in 2009.[2] This is believed to be the best researched, best controlled, most thorough study ever made of trace currents and temperatures. The document is over 90 pages long and contains over 75 charts and tables. The results for external traces are (in our opinion) more evolutionary than revolutionary. The IPC-2152 data result in currents approximately 25% lower than those shown in the original IPC-2221 set of curves, and they are much more detailed and complete. But the results for the internal traces are revolutionary. It turns out the internal traces cool *better* than do the external traces. That is because it turns out the board materials conduct heat away from the trace better than the air does. This is the one assumption the original researchers got very wrong.

I (Brooks) wrote my first article on the relationship between PCB trace currents and temperature in 1998.[8] In that article I compared two data sources: the then-current IPC curves and some data from a 1968 issue *of Design News* magazine. There was about a 40% difference in the results between those two sources that he was not able to explain.

Through the years I began to speculate that the difference might be related to thermal gradients on the traces. In the meantime, the IPC-2152

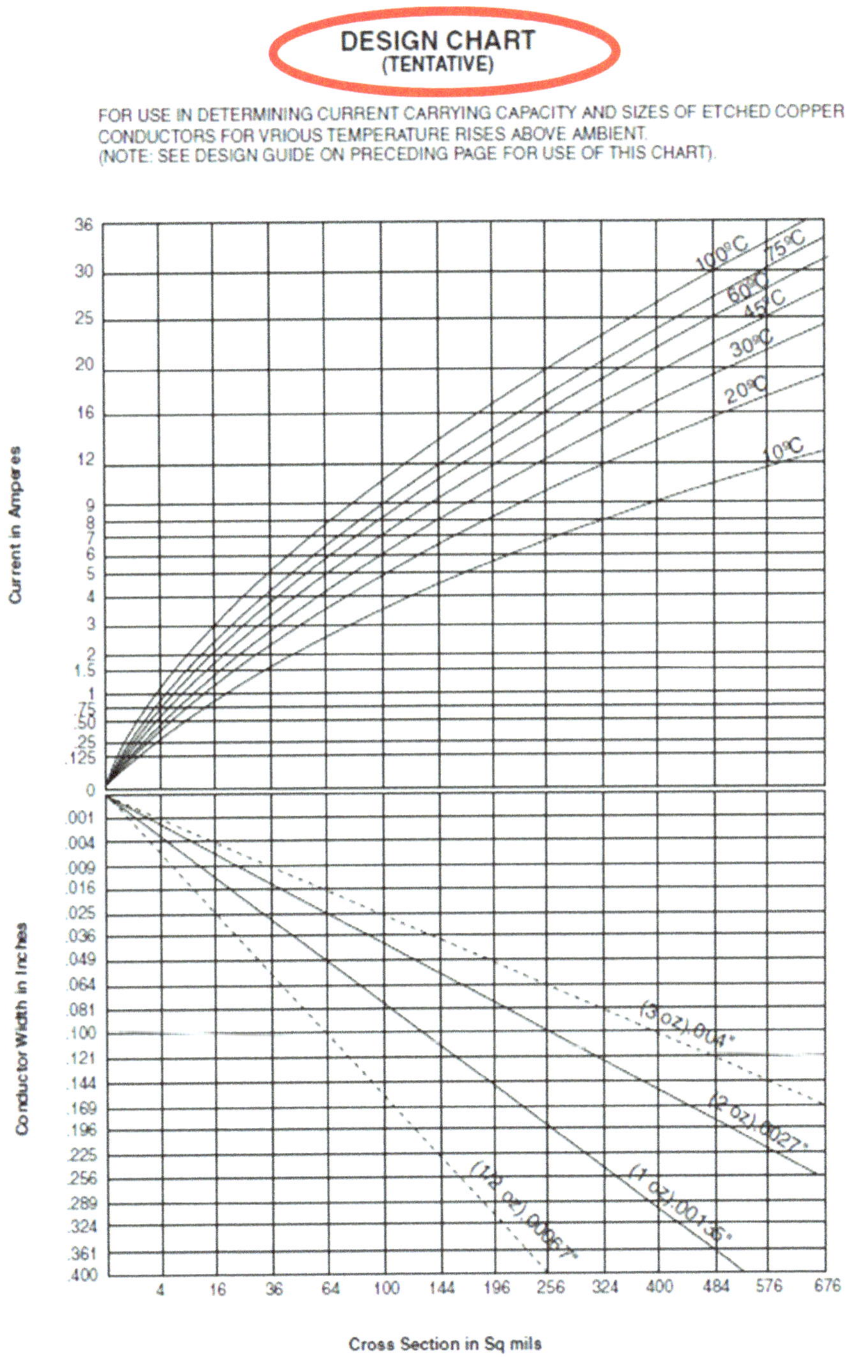

Figure 1.1 First known chart of trace current/temperature relationships. Note the word "Tentative" in the title. (Source: NBS Report 4283.)

standard was published, suggesting an even bigger difference between the IPC data and the 1968 *Design News* data.

Then, in late 2014, I opened up a dialog with Johannes Adam, in Leimen Germany. Johannes had authored a thermal simulation program called Thermal Risk Management (TRM), and I reasoned that that simulation package might finally help me answer the question about thermal gradients on traces. Johannes graciously offered me a license for TRM, and we began a collaboration that has lasted to the present time.

Our first objective was to determine if TRM could accurately model the IPC-2152 results. When this was accomplished (Chapter 6), we began looking into sensitivity analyses; that is, what happens if we change board material or layout parameters (Chapter 7)?

Those results were so encouraging that we decided to try to use TRM to look at the thermal properties of vias (Chapter 8). The results of those simulations were so unexpected and dramatic that our collaboration began to change in scope. First, we knew that we needed to experimentally validate those via simulations. Prototron Circuits was very helpful in us doing that. But then we realized that we needed to learn a lot more about the properties of the test boards we obtained (properties discussed in Chapters 2 and 3.) Appendixes A and B resulted from those efforts.

Those investigations led to our fusing simulations and experiments (Chapters 7 and 8 and Appendix G.)

And, finally, the idea for this book was born. Almost all the issues related to PCB trace temperatures are combined into a single publication. We start with the basic theory of trace heating and cooling. This includes resistivity and how to measure it, and the thermal cooling capacity of board material and how to measure that. Then we look at how the IPC curves were developed, what they mean, and how they might be simulated with computer software. Then we look at sensitivity analyses; how many different variables are there and what happens when they are varied. This includes looking at the dynamics affecting via temperatures. And then we look at what happens when we catastrophically overload a trace. One thing happens if we overload it greatly (and it melts within seconds); something else happens if we slightly overload it and it takes from 30 minutes to over an hour to fail. We provide experimental data to illustrate the various results presented in the chapters.

We have tried to make the information this book thorough and complete. Of course that is not possible. This is a very complex topic and there are a great many variables (and not a few uncertainties) involved. We have tried to open a door with this book, but there is still a lot to be discovered on the other side of that door.

We believe it is not possible to exactly characterize a board design without actually testing the physical product. And even then, there is likely to be a lack of repeatability from one product to another. And we also believe it is becoming impractical to get close enough to the final results with charts, graphs, and equations. For a really optimal analysis of a board design a more complex simulation technique is required, one that requires some healthy computer analysis. And, as indicated in the preface, there is precedent for this in our industry. In the 1990s it became important (in some designs) to design impedance-controlled traces, and formulas were developed to support this need. But now we know that those formulas are not accurate enough; field-solver techniques are usually required. The same thing is happening related to trace temperatures. TRM is one such approach to this need.

1.3 A Note about Consistency

We detailed above how topics of investigation led from one to another. We were perhaps 2 years into our collaboration before we began seriously thinking in terms of a book. Unfortunately, some of our simulations and experiments had been concluded by then and some of the resources we had were no longer available. That means simulations and experiments were performed at different points in time with little thought given to consistency in models. Thus the reader will note that models and experiments in one section might have slightly different parameters than they have in other sections.

In addition, the industry is careless about how it estimates the nominal thickness of a trace. For example, the thickness of a half-ounce trace (we define that in Footnote 1, Chapter 5) is often given as either 0.60 or 0.65 mil. That converts to a one-ounce trace being 1.2 to 1.3 mil thick. The corresponding mm equivalents would be 0.015 to 0.017 for half-ounce traces and 0.03 to 0.034 for one-ounce traces. One-ounce traces are often given as 0.035 mm, converting back to 1.38 mil! On top of this, the IPC puts a 10% tolerance on trace thickness.

Finally, conversions between imperial and metric dimensions are not clean. For example, a 250-mil trace converts to 6.35 mm, often rounded to 6.3 or even 6.0 for convenience. And, as mentioned above, a one-ounce thickness is often expressed as 0.035 mm (1.38 mils) or 1.35 mils (0.034 mm).

If all this is bothersome to the reader, you can treat each section as an independent topic unless we specifically refer back to a previous section. The minor differences in consistency have no effect on the overall conclusions from the various discussions.

End Notes

1. National Bureau of Standards Report 4283, *Characterization of Metal-Insulator Laminates,* 1956 (no longer in print). This image appears on page 26.

2. For more information, see IPC-2152, Standard for Determining Current Carrying Capacity in Printed Board Design, IPC, August, 2009, Appendix A.7, p. 85. A copy of the original NBS chart is included there as Figure A-89, p. 86.

3. MIL-STD-1495, Multilayer Printed Wiring Boards for Electronic Equipment (currently out of print).

4. MIL-STD-275, Printed Wiring for Electronic Equipment (currently out of print).

5. ANSI/IPC-D-275, Design Standard for Rigid Printed Boards and Rigid Printed Board Assemblies, Figure 3-4, IPC, September 1991, p. 10.

6. IPC-2221, Generic Standard on Printed Board Design, 1998, superseded by IPC-2221A, Generic Standard on Printed Board Design, May 2003, IPC.org, Figure 6-4, p. 41.

7. http://www.ultracad.com/thermal/thermal.htm.

8. Brooks, D., "Temperature Rise in PCB Traces," *Proceedings of the PCB Design Conference West,* Miller Freeman, Inc., March 23–27, 1998.

CHAPTER

2

Contents

Materials Used in PCBs

2.1 Bottom Line

- PC boards consist of copper laminated onto or plated over a dielectric.
- For trace current/temperature relationships the most important parameters are copper resistivity, dielectric thermal conductivity coefficients, and uniform copper thickness.

2.2 Background

At first glance, a PCB is a pretty simple thing. There are one or more copper layers etched into conducting patterns, and one or more substrate layers to hold the copper. For most applications up until around the 1990s, that concept was sufficient. Board design was a drafting exercise often referred to as "connecting the dots." But then as circuit and component rise times got faster, individual traces began to act more like components, and problems like transmission line impedance control and EMI began to surface. This opened up a whole new set of problems for some designers, especially those who had no background in basic electronics. This trend developed into a topic we now refer to as "signal integrity" related to board design.

As circuit components became faster and more complex, the power to operate them also began to increase. And this generated heat. Temperature and thermal management became a significant concern in some designs. More and more designers began to rely on standards like IPC-2152 to help determine how to appropriately size traces.

But stepping back for a moment, we might ask, Why does a trace heat up in the first place? Why does trace size matter? After all, copper has a very low resistance to current, almost zero. And it conducts heat very well. Why isn't that sufficient?

In Chapters 2 through 4 we will look first at copper itself, and answer the question of whether copper is just copper. That is, are there any relevant differences between types of copper used on a PCB? We will also look at the types of laminates PCBs are made from, and what parameters are important to us. Then, we examine why traces heat up and how they cool. Where is the balance between heating and cooling? How do we measure trace temperature? And finally, is there anything that can change the balance between heating and cooling in the sense that a seemingly stable temperature becomes unstable and thermal runaway occurs?

2.3 Copper Used in PCBs

2.3.1 Copper-clad laminates

For surfaces where there will be conducting traces, the board fabricator typically purchases basic board material covered with a copper foil from one or more suppliers. The copper foil is typically either electrodeposited (ED) or it is rolled. In the ED foil manufacturing process, the copper is deposited on a conducting drum (which acts as a cathode) that slowly rotates in a copper sulfate solution. The drum is connected to a negative DC power supply and the positive supply side is connected to an anode in the solution. The slower the drum rotates, the thicker the plated copper foil (Figure 2.1).

Rolled copper starts with a billet of copper alloy that is rolled between two rollers. This flattens the copper somewhat. It then goes through another pair of rollers, which flattens it further. Successive rolling operations result in the copper becoming thinner and thinner until it reaches the desired thickness (Figure 2.2).

In both cases, the result is a sheet of thin copper foil. But there are some significant differences in these processes that may be of importance to the board designer. Rolled copper is flatter (depending on the roller surfaces)

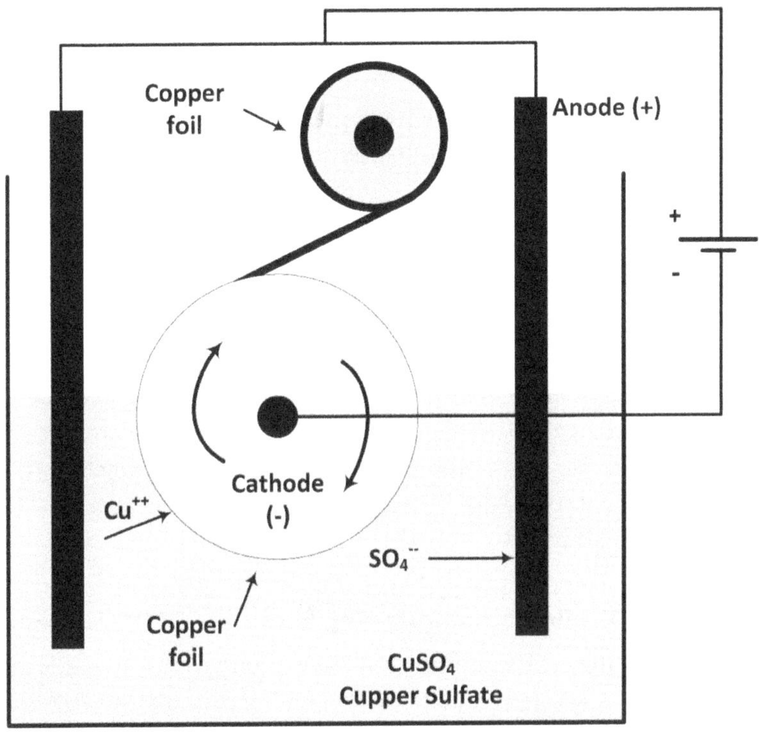

Figure 2.1 ED foil process.

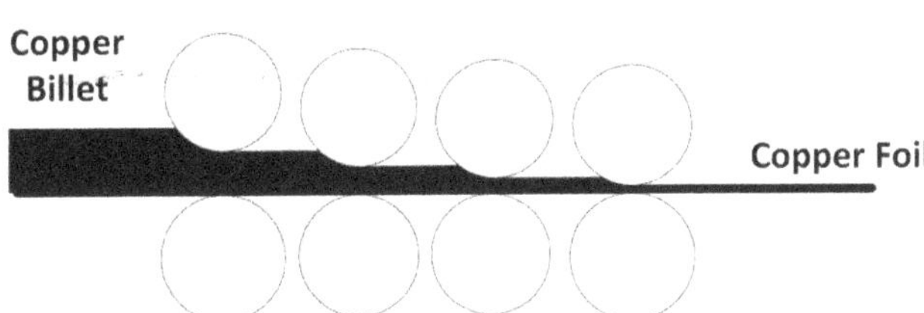

Figure 2.2 Rolled copper foil process.

than ED copper. That is, the surface roughness is greater for ED copper. Rolled copper is also mechanically more robust than the pure ED copper. But the ED copper has two surfaces, and the drum side surface is smoother than the outside surface so the outside (rough) surface provides a good surface for the copper to bond to the laminate without any intermediate processing. Rolled copper may require an intermediate step to rough up the surface in order to provide a good bond to the laminate. The ED copper can also produce thinner foils than can the rolled copper. The practical importance of these differences are[1]

- Rolled copper is better for high-frequency applications (where surface roughness may have signal integrity implications);
- ED copper is better for bonded assemblies;
- ED copper has a lower resistivity (after bonding to a laminate) (1.62–1.66 μohm-cm) than rolled copper (1.74–1.78 μohm-cm).

The copper foil process is pretty well controlled and tolerances are good. Foil thickness is typically held within 10% or better. Resistivity is reasonably constant across the surface of the foil. While there is opportunity for contamination to occur between the foil and the board (which may impact some thermal properties), this should be limited to a few spots on the board and should not significantly impact the thermal performance of the board.

2.3.2 Copper Plating Manufacturing Step

The board fabricator starts with the basic board material and builds up a finished board. Inner layers will go through etching processes to etch the trace layer pattern into the copper, and for any blind and/or buried vias, may go through a separate plating step. Then the layers are assembled and bonded together. There is a point where the layers of the board have been fully built up, and the board (really manufacturing panel) exists as an assembly with solid copper foil on the top and the bottom and holes drilled for the vias.

From here, the process can take one of two directions:

1. The top and bottom patterns can be etched into the copper, and then board (and traces) plated with copper to cover the walls of the vias (this process is referred to as pattern plating);
2. The entire board area (top and bottom) can be plated with copper and then the trace patterns etched into the combined foil plus plating layer (referred to as panel plating).

In both cases the outside layer copper foil is covered with plated copper. This plating step is very similar to the plating step for copper foil—the board is dipped into a copper solution and the copper is electroplated onto the board. Then the board is withdrawn from the solution (Figure 2.3). In the first example the trace patterns are etched before plating, and in the second example the pattern is etched after plating. In both cases the traces consist of copper foil covered with plated copper. There may be minor differences

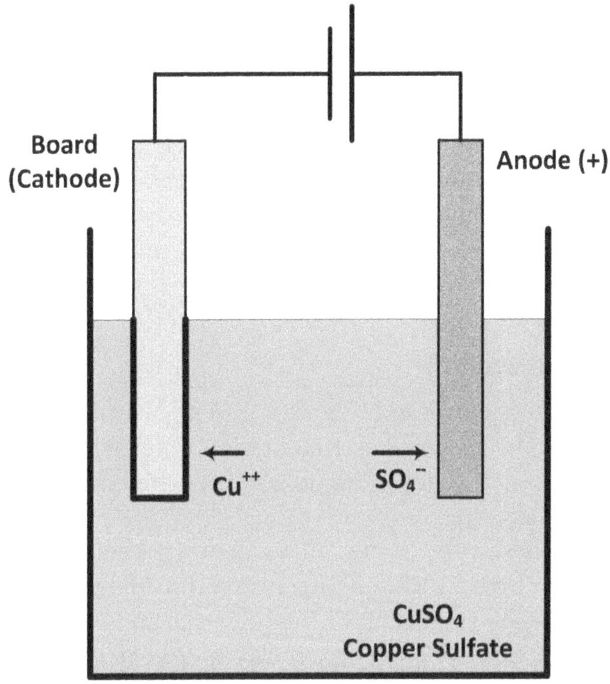

Figure 2.3 Plating of board.

in the thickness and shape of traces etched in these two manners, and these differences might be important to signal integrity engineers; they may have an impact in trace impedance calculations for example. But they are unlikely to impact the trace's thermal characteristics.

But there are other things here that might be important to us. First, and perhaps most important, is that this plating process itself is difficult to control. Different areas of the board may plate to significantly different thicknesses. Thickness variation of up to 50% should not be unexpected and there is anecdotal evidence on the internet that plating thickness might vary by as much as 100% around the board. This variation can be controlled somewhat by the fabricator, but that might involve additional steps, modification to the board design, and additional expense. Such efforts might reduce the variations in thickness, but variations as much as 20% may still occur. As we will see below, trace thickness has a significant impact on trace resistance, which therefore has an impact on trace heating.

In Chapter 13 we will show that traces do not heat uniformly. A major reason for this lack of uniformity is variations in trace thickness.

Second, plating may impact resistivity (see the next section).

Third, the many processing steps open up the possibility of contamination, especially between the foil and plated copper layers. Such contamination, if it exists, may impact the thermal characteristics of the board, especially at elevated temperatures.

Finally, the pattern etching process has a tolerance that applies to the finished width of the trace. If the trace width varies along the trace, this may impact the thermal characteristics of the board, and if it does, where hot spots may exist along a trace. The narrower a trace, the more important this tolerance is.

2.3.3 Copper Resistivity

The resistivity of copper is published in a very large number of places.[2] Resistivity is typically specified as 1.68 μohm-cm for pure copper and 1.72 μohm-cm for annealed copper. The problem is complicated by the number of copper alloys there are.

There are hundreds of copper alloys defined. There is a classification system called the Unified Numbering System for Metals and Alloys (UNS for short). Copper alloys have UNS numbers starting with Cxxxxx.[3] Coppers (UNS C10100 to C15999) have a purity => 99.3%, and high-copper alloys (UNS C16000 to C19999) have a purity > 96%. Selecting random UNS alloys in end note 3 in the range of C10100 to C14500 results in a range of resistivities from 1.69 to 1.86 μohm-cm.

If there are any impurities in the copper, the resistivity will change. An interesting chart in a paper titled "High Conductivity Copper for Electrical Engineering"[4] shows the approximate effect of impurities on mass electrical resistivity (mass resistivity = volume resistivity * density). Changes in resistivity depend very much on the impurity and the concentration, but the variation can be 25% with even small changes of some impurities. (See Chapter 3 for more on copper resistivities.)

2.3.4 Summary

For trace current/temperature considerations, the following copper trace parameters are important:

1. Variations in finished trace thickness;
2. Variations in resistivity;
3. Points where contamination may exist between the foil and the board material or between the plated copper and copper foil parts of a trace.

2.4 Dielectrics Used in PCBs

A search of "PCB dielectrics" on the internet results in a significant variety of laminates a board designer might choose from. The most common dielectric is standard FR4, but even that category has many variants. Recent new material choices have arisen because of signal integrity needs. High-frequency materials have been developed for their dielectric coefficient stability with frequency and/or to minimize their dielectric losses. But none of that matters for our purposes. There are four parameters that matter from a current/temperature standpoint, and only one of those matters at normal temperatures.

2.4.1 Thermal Conductivity (Tcon or k)

Dielectrics conduct heat in two directions. One is in the *x-y* plane, parallel to the surface layer, and therefore parallel to the traces. This is sometimes called the in-plane conductivity. The other is in the *z*-axis, or perpendicular to the surface. This is sometimes called through-plane. The measure that reflects this thermal conductivity is called the thermal conductivity coefficient, sometimes referred to as Tcon-x and Tcon-z, respectively. Typical values for Tcon-x might be 0.5 or 0.6 and for Tcon-z might be 0.3 to 0.5 (compare with copper at 386). Material specifications frequently fail to include Tcon values, and if they do they are unlikely to specify whether the value applies to the in-plane or through-plane direction. And different measurement procedures used to measure Tcon may produce differing results. Units of Tcon are watts/meter-degree Kelvin, or W/mK.

There are now tools available to measure thermal conductivity at a relatively modest cost. One such tool is offered by C-Therm Technologies in New Brunswick, Canada (see Appendix A).

This parameter is important to us because it directly relates to the dielectric's ability to conduct heat away from the trace.

Three other material parameters are important to us if the trace reaches a moderately high temperature (say higher than 100°C).

2.4.2 Glass Transition Temperature (Tg)

See IPC Test Method 2.4.24)[5] for more details. Glass transition temperature is where the polymer board material transitions from a hard, glasslike material to a soft, rubbery material. (This is *not* to be confused with melting.) It is really a range of temperature over which this transition occurs, and Tg is defined as the center of that temperature range. During the transition, the board material coefficient of thermal expansion (CTE) increases, which means the board itself expands. Since the board material is somewhat constrained by copper traces and planes in the *x-y* direction, the dielectric expansion tends to occur in the *z*-axis, sometimes putting mechanical stress on nearby vias. In extreme cases, via conductivity may be lost as a result of this expansion.

Since the dielectric material is changing form, and since the dimensions are changing, therefore the thermal conductivity is also changing. This impacts the cooling characteristic of the material. A board can be perfectly usable at temperatures above Tg, and the effects of Tg are reversible.

2.4.3 Decomposition Temperature (Td)

Decomposition temperature is the temperature at which a PCB material chemically decomposes (i.e., the material loses at least 5% of mass). This results in changes in material properties and therefore in a change in which the material can conduct heat away from the trace. Td is important because the changes that result are irreversible. (This is a consideration for assembly processes, especially flow soldering operations.)

Fortunately, typical Td specifications are relatively high, often in the range of 350°C.

2.4.4 Time to Delamination (T260/T288)

There is a test known as Time to Delamination and is defined in IPC Test Method 2.4.24.1.[6] It is sometimes referred to on specification sheets as Thermal Resistance. A test sample is incrementally raised in temperature 10°C/min to 260°C (or 288°C) and then is held at that temperature. When a destructive event occurs (such as delamination, cracking, moisture release, stress relaxation, decomposition or a sudden movement) a sensor detects the change. The time in minutes to the event at 260°C (or 288°C) is the time to delamination. The delamination event is destructive and irreversible. Times for different laminates can vary between a few minutes to an hour or more.

This is relevant for us in the following situation. Suppose we have a trace carrying a very high current that needs supplemental cooling (forced air flow or heat sink). Then suppose something happens that for some reason degrades the supplemental cooling and the trace temperature rises to these temperatures. The temperature may hold steady for a while, but if the board begins to delaminate, all bets are off. Once delamination starts, the conductive cooling through the board material is disrupted and the temperature will start increasing, often resulting in a thermal runaway condition and subsequent melting of the conductor (see Chapter 12).

2.4.5 Summary

For trace current/temperature considerations, the following laminate material properties are important:

1. Thermal conductivity (Tcon or k): relates to how well the material carries heat away from the trace.
2. Glass transition temperature (Tg): temperature at which the polymers change from hard, glassy-like appearance to softer plasticlike appearance (reversible).
3. Decomposition temperature (Td): temperature at which the board material begins to decompose (irreversible).
4. Time to delamination (T260/280): time for the board to begin delaminating (irreversible and very destructive). This event can lead to thermal runaway.

End Notes

1. "Copper Foils for High Frequency Materials," Rogers Corporation, Chandler, AZ, 2015.
2. See, for example, http://hyperphysics.phy-astr.gsu.edu/hbase/tables/rstiv.html.
3. See, for example, http://www.matweb.com/search/SearchUNS.aspx.
4. Chapman, D., *High Conductivity Copper for Electrical Engineering,* Copper Development Association Publication No. 122, European Copper Institute Publication No Cu0232, originally published May 1998, Figure 1, p. 6. Paper can be downloaded at http://copperalliance.it/uploads/2018/02/pub-122-hicon-copper-for-electrical-engineering.pdf.
5. See IPC Test Manual IPC-TM-260 Test Method 2.4.24, *Glass Transition Temperature and Z-Axis Thermal Expansion by TMA.* A complete set of test manuals and test methods can be freely downloaded from http://www.ipc.org/test-methods.aspx.
6. See IPC Test Manual IPC-TM-260 Test Method 2.4.24.1, *Time to Delamination (TMA Method).* A complete set of test manuals and test methods can be freely downloaded from http://www.ipc.org/test-methods.aspx.

CHAPTER

3

Contents

Resistivity and Resistance

3.1 Bottom Line

- Resistivity is an inherent property of a chemical element. For our purpose and interest, it is a point property. That is, it can vary from point to point along a trace. It also varies with temperature.
- Silver, copper, and gold have the lowest resistivities of any elements known.
- Resistance per-unit-length is resistivity divided by area, and is also a point parameter that varies with temperature.
- All this will be relevant for us when we look at trace temperature changes in response to changing current.

3.2 Resistivity

Resistivity is a property of a chemical element that is related to the resistance of that element to the flow of current. Its units are Ω-unit length. All elements have resistivity. In order for us to understand resistivity (and resistance), let us first understand the nature of current. To do that, we must look at atomic structure. Figure 3.1 is a simple model of an atom. An atom has a nucleus consisting of protons and neutrons, surrounded by electrons. (We know that atoms are much more complicated than this, but the following simplified model provides all the insight we need for now.) The number of protons defines the element. For example, copper has 29 protons. An atom has the same number of electrons (negatively charged) as it has protons (positively charged). So in all but very special cases, atoms are electrically neutral.

Electrons fly around the nucleus in energy shells. We don't know how to draw energy shells, so we draw them for convenience as concentric circles. Thus, the model looks like a sun with planets orbiting around it, giving it the name "planetary model." Now these shells have maximum capacities. The inner shell has a maximum capacity of two electrons. The next shell has a maximum capacity of 8 electrons. It tends to get more complicated from there, but it is sufficient to understand that much.[1]

Now here is the punch line: Current is the flow, or movement, of electrons. (One amp of current is the movement of 6.25×10^{18} electrons passing across a surface in one second. That's a lot of electrons!) Atoms with a single electron in their outermost shell tend to have the lowest resistance to current flow. The reason is that that electron is not very tightly coupled to the nucleus. If there are more electrons in the highest energy shell, they tend to be more tightly coupled to the nucleus. Loosely held electrons are the easiest to displace. So if we apply a force (think voltage) across a copper conductor, that force can dislodge the electrons in the outermost shell from one atom and shift it further along the conductor.

Figure 3.2 is a model of a copper trace. Imagine a voltage (force) applied across the conductor by a battery. The battery pushes an electron onto the trace. It is attracted by the opposite charge at the far end of the trace. An

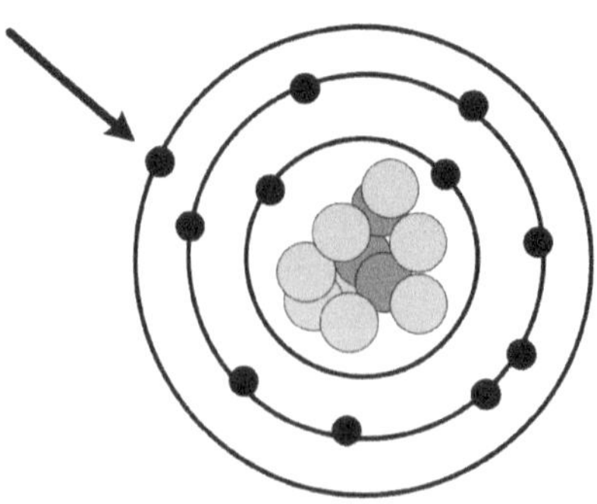

Figure 3.1 Planetary model of an atom.

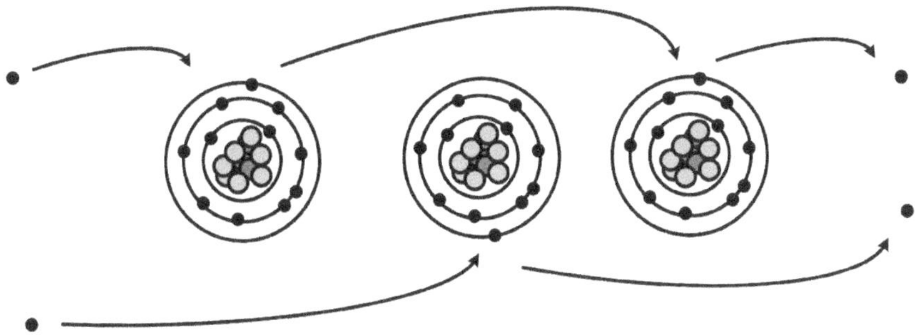

Figure 3.2 Model of a conductor.

electron must therefore leave the trace at the opposite end because the trace (and the atoms making it up) must remain electrically neutral.[2] That movement (shift) in electrons is the definition of current. The injected electron travels along the trace, but in doing so it collides with an electron that is already there. The injected electron may stay in place, replacing the electron it collided with, and if so the displaced electron continues forward under the force of the voltage. This forward motion of the electrons, and their collisions, continues at a very high rate.[3]

Now since the single electrons in the outer shell are loosely held, it does not take much energy to displace them. But the required energy is not zero. There is a small amount of energy that is dissipated each time one electron collides with another and the second electron is displaced. That energy loss is manifested in increased heat. And it is that energy loss that is the nature of resistance and resistivity.[4]

Now we intuitively know that silver, copper, and gold are excellent conductors of electricity. They have three characteristics of particular interest:

1. They have the lowest values of resistivity of all elements;
2. They all have a single electron in their outer shell;
3. They are all solids at room temperature.

Their reported resistivities (at 20°C) are:[5]

- Silver: 1.6×10^{-8} ohm-m = .63 μohm-in;
- Copper: 1.7×10^{-8} Ohm-m = .67 μohm-in;
- Gold: 2.2×10^{-8} Ohm-m = .87 μohm-in.

Note that units of resistivity are ohm-length.

For comparison, silicon has a resistivity of 6.4×10^2 ohm-m and has four electrons in its outer shell. It is known as a semiconductor. Glass has a resistivity of 1.0×10^9 to 1.0×10^{15} ohm-m. Glass is an excellent insulator.

The resistivity of copper reported above is the resistivity of pure copper. If there are any impurities present, those impurities will have higher resistivities. Collisions with their electrons will dissipate more energy and the overall resistivity will increase. So any contamination of the copper used in a trace will increase the resistivity of the copper and the resistance of the trace.

3.3 Resistance

Resistivity is a property associated with an element (such as copper) or a combination of elements (such as a copper alloy). Resistance is a parameter of an element that has a specific shape (such as a conductor). The resistance of a conductor is found by first dividing the resistivity by the cross-sectional area of the conductor. That results in units of Ω/unit-length. We then multiply this by the length to calculate the resistance. Those two steps are combined in (3.1):

$$R = (\rho / A)^* L \tag{3.1}$$

where

R = resistance in ohms

ρ = resistivity in ohms-length

A = cross-sectional area in square units

L = length in units

So if we know the resistivity of copper (say 0.67 μohm-in), then we would calculate the resistance of a 0.5 oz (0.00065 in) 100 mil (0.1 in) wide, 6″ long trace as

$$R = ((0.67^*10^{-6})/(.00065^*0.1))^*6 = 0.0618 \text{ ohms} \tag{3.1a}$$

Conversely, if we know the resistance of a trace (because we measured it directly or calculated it from Ohm's law), then we can calculate the resistivity of the trace material by

$$\rho = R * A/L \tag{3.2}$$

Resistivity and resistance are point concepts in the sense that they can vary from point to point along a trace. Resistivity can vary, perhaps because of impurities in the copper. Resistance can vary because resistivity is varying, or because the conductor dimensions are varying.

3.4 Thermal Coefficient of Resistivity (α)

Resistivity also increases with temperature. The reason is that temperature is motion. That is, as the temperature increases, atomic and molecular motion increases. (Absolute zero, –273°C, is defined as the point where all atomic motion ceases.) When the atomic motion increases, there are more collisions among the electrons carrying the current, resulting in an additional loss of energy.

Since resistivity changes with temperature, it is important (when specifying resistivity) to add the specific temperature reference related to the resistivity value being specified. So, the resistivities of silver, copper, and gold stated above are specified to the be the values at 20°C.

The change of the resistivity of copper (indeed almost any element) with temperature is predictable. The property, *thermal coefficient of resistivity* (α), is the relative change in resistivity associated with a given change in temperature (3.3). Thermal coefficients of resistivity for most elements are readily available on the internet.

$$\alpha = \frac{\Delta\rho / \rho}{\Delta T} \tag{3.3}$$

where:

α = Thermal coefficient of resistivity

ρ = Resistivity

$\Delta\rho$ = Change in resistivity

ΔT = Change in temperature

Since resistance is simply resistivity modified by the geometry of the conductor, it follows that we can substitute resistance for ρ in (3.3) and get (3.4):

$$\alpha = \frac{\Delta R / R}{\Delta T} \tag{3.4}$$

α itself is a function of temperature. It is correctly stated as a value at some specific temperature (most commonly 20°C). So, (3.4) leads to the fundamental relationship between resistance and temperature:

$$R(T) = R(T_o) \times (1 + \alpha_o{*}(T - T_o)) \tag{3.5}$$

where

T = Temperature of interest

T_o = Reference temperature

$R(T)$ = Resistance at the temperature of interest

$R(T_o)$ = Resistance at the reference temperature

α_o = Thermal coefficient of resistivity at the reference temperature

Equation (3.5) applies equally well to resistivity, ρ:

$$\rho(T) = \rho(T_o){*}(1 + \alpha_o{*}(\mathrm{T} - T_o)) \tag{3.6}$$

where

T = Temperature of interest

T_o = Reference temperature

$\rho(T)$ = Resistivity at the temperature of interest

$\rho(T_o)$ = Resistivity at the reference temperature

α_0 = Thermal coefficient of resistivity at the reference temperature

Matula[6] has a table of copper resistivities as a function of temperature. The figures are for at least 99.99% pure copper. This goes up to and including the melting point of copper and then for liquid copper (see Figure 3.3). From his data we can derive the implied thermal coefficient of resistivity using Equation (3.3). Table 3.1 is derived from Matula's data.

3.5 Measuring Resistivity

Suppose we wish to measure the resistivity of a copper trace. We can do so relying on (3.2). But there are some significant barriers to doing so.

First, we need to determine R, the resistance of the trace. We can do so a couple of different ways. One is to measure it directly with an ohmmeter. There are precision meters available that can do this with a high enough

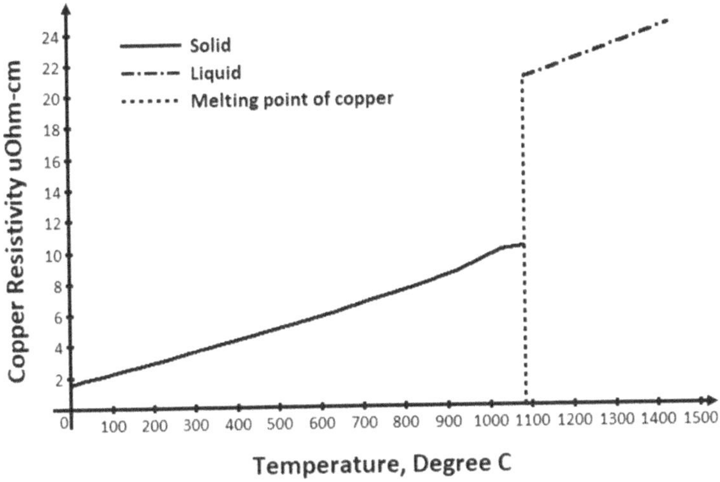

Figure 3.3 Curve of copper resistivity vs. temperature.

Table 3.1
Resistivity as a Function of Temperature

Temp (°C)	Resistivity μohm-cm	Implied α from 20°C
0	1.541	—
20	1.676	—
27	1.723	.0040
77	2.061	.0040
127	2.400	.0040
227	3.088	.0041
327	3.790	.0041
427	4.512	.0042
527	5.260	.0042
627	6.039	.0043
727	6.856	.0044
827	7.715	.0045
927	8.624	.0046
1027	9.950	.0047
1084	10.17	.0048
1084*	21.01	—
1127*	21.43	—
1227*	22.42	—
1327*	23.42	—
1427*	24.41	—

* = Liquid copper.

precision, or we can apply a known current through the trace and measure the voltage (or vice versa) and calculate resistance using Ohm's law, $R = V/I$. This can also be accomplished with a reasonably high precision, especially under controlled conditions.

Second, we need to be able to measure or calculate the cross-sectional area of the trace. This means knowing (or measuring) the trace width and thickness. It is important to recognize that *this cannot be done by using the nominal dimensions of the design*. There is far too much tolerance in the dimensions to be able to do this. If there is copper plating applied over copper film, trace thickness can vary from nominal by as much as 50% (or even more). The photo-etching process can have a tolerance of at least 1 or 2 mils.

Third, the trace length would seem to be measureable, but there are possible problems here, too. If there are pads at the ends of the trace, their area may need to be considered, especially if the pad is approximately the same size as the trace. Only in the case where the pad is significantly larger than the trace (in which case its resistance is significantly lower than that of the trace) can we perhaps be able to ignore the pad. If the pad's dimensions are different than that of the trace, the calculations may become significantly more difficult.

Finally, the temperature may be a factor. If the trace is at an elevated temperature and we wish to end up with the resistivity at 20°C, then we need to do the type of conversion related to (3.5). This implies we know *both* the elevated temperature and α, neither of which can be casually assumed.

Often, when we want to know trace dimensions precisely, we microsection a trace to take the measurements. Figure 3.4 shows a microsection from a test board. The view on the left is the full microsection showing a 1-inch long section of a trace (top and bottom layer). The view on the right expands part of the view.[7] The problem with microsectioning is that (a) it

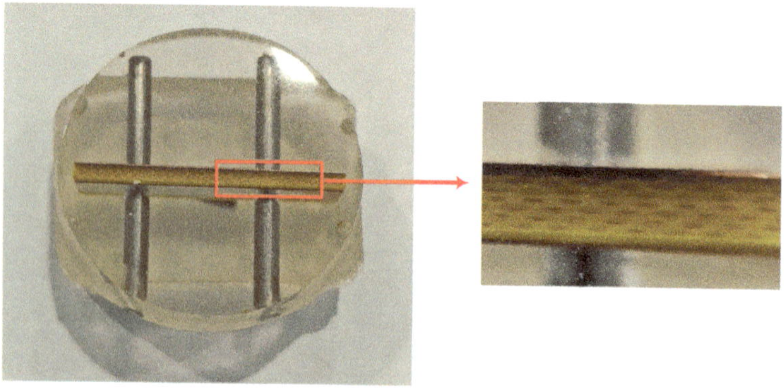

Figure 3.4 Typical microsection sample from a test board.

is destructive, and (b) the typical microsection measurement tolerance is around 0.1 mil. That is about a 10% or 15% measurement error for a half-ounce trace! Furthermore, microsectioning will give you the dimensions of the trace *at that point*. But the dimensions might change as we move along the trace (see more about this in Chapters 7 and 13). Microsectioning cannot help us determine the extent of such variation.

It is our opinion that with today's technologies, it is impractical to try to measure the resistivity of a typical PCB trace without the use of advanced and sophisticated measurement tools in a well-controlled lab. We simply cannot do it precisely enough. There are numerous anecdotal examples reported on the internet of people trying to measure resistivity in numerous ways, with results ranging from 1.6 to 2.7 μohm-cm, but it is believed all of them suffer from the uncertainties outlined above.

3.5.1 Resistivity Investigation

With the help of a local PCB fabrication house,[8] a special test board was prepared to measure the resistivity of copper foil, plated copper, and their combination. The special design of the test board enabled the calculation of trace resistivity with reasonable precision. It was determined that the resistivity of the copper plating was almost exactly that of pure copper—just what one might expect. The details of that investigation are reported in Appendix B.

3.5.2 Nondestructive Measurements

There is an emerging technology, computed tomography ((CT) industrial scanning) that might make nondestructive trace measurements more practical. The resolution is not quite high enough today to be useful in these types of analyses, but we are optimistic that it might be in the not-too-distant future.

We had the opportunity to send a section of a board to Jesse Garant Metrology Center, Windsor, Ontario, for x-raying. The results of that investigation are reported in Chapter 16.

End Notes

1. For those readers who are more comfortable thinking about conduction bands filled with a sea of free electrons, consider this: That is also a model (the valence band model). There is no "sea" and the electrons are not totally "free." Take any volume somehow defined in a copper conductor. No matter how (small) you define that volume, it must remain charge-neutral. Otherwise, charges will simply shift to maintain neutrality. In moving electrons through

that sea, some energy is lost. It takes very little energy per electron to do so, but multiply that energy by 6.25×10^{18} electrons and the numbers start to add up.

2. From this much we can infer two of the most important laws in electronics: (a) current must flow in a loop, and (b) current is constant everywhere in the loop. See Brooks, Douglas G., *PCB Currents, How They Flow, How They React*, Prentice Hall, 2013, Chapters 1 and 4.

3. Some students get confused by this model and how it squares with the fact that signals propagate at the speed of light. There are *two* velocities here. One is the speed of a single, individual electron. It travels at a speed called the drift velocity—most of us can walk faster than that. The other is the speed with which the transfer of energy takes place among the atoms that make up the trace. That transfer of energy happens at the speed of light. Again, see note 2, above.

4. In our context within this book, the flow of electrons is related to the current, I, the loss of energy is related to the resistance, R, and the resulting heat is related to the I^2R power dissipation that we will discuss in the following chapters.

5. Resistivity values for all the elements are readily available on the internet and in various handbooks.

6. Matula, R. A., Purdue University, "Electrical Resistivity of Copper, Gold, Palladium, and Silver," as reprinted in *J. Phys. Chem. Ref. Data*, Vol. 8, No. 4, 1979. See especially Table 2, p. 1161.

7. See more about microsectioning, and another image, in Chapter 16.

8. Prototron Circuits, Tucson, AZ, graciously provided some test boards for this investigation.

CHAPTER

4

Contents

Trace Heating and Cooling

4.1 Bottom Line

- Traces heat because of I^2R power dissipation. They cool because of conduction through the dielectric, convection through the air, and radiation.
- A stable temperature is reached when heating equals cooling.
- However, temperature is a point concept; temperatures are rarely uniform along a trace.

4.2 Overview

Traces heat because of I^2R power dissipation, increasing the temperature above the ambient temperature or initial condition of the trace.[1] The trace's resistance is inversely proportional to the trace's cross-sectional area, so larger traces do not heat up as much as do smaller traces. Traces cool because of conduction, convection, and radiation, lowering the temperature. A trace's cooling characteristics are related to the surface area of the trace at any point along the trace. If the power is changing with time, then the total energy being dissipated in the trace will be changing with time. The cooling may also change with time. A stable temperature is reached when heating equals cooling over a stable time frame (Figure 4.1).

4.3 Trace Heating

We noted in Chapter 3 that traces have resistance—not much, but resistance is not zero. Therefore, a current through a trace will result in a voltage across the trace (and vice versa). Power is voltage times current:

$$\text{Power} = V * I \text{ (units are watts)} \tag{4.1}$$

By Ohm's law, $V = I * R$, so we can, by substitution, rewrite (4.1) as

$$\text{Power} = I^2R \text{ (units are watts)} \tag{4.2}$$

We often take the term "power" very casually, so let's take a moment to be clear about the difference between *power* and *energy*.

4.3.1 Power and Energy

Energy is the amount of work done. Power is the rate at which we do it. A 100-watt light bulb dissipates power at the *rate* of 100 watts (per unit time). The total amount of *energy* it dissipates in an hour is (100 watts * 1.0 hour =)

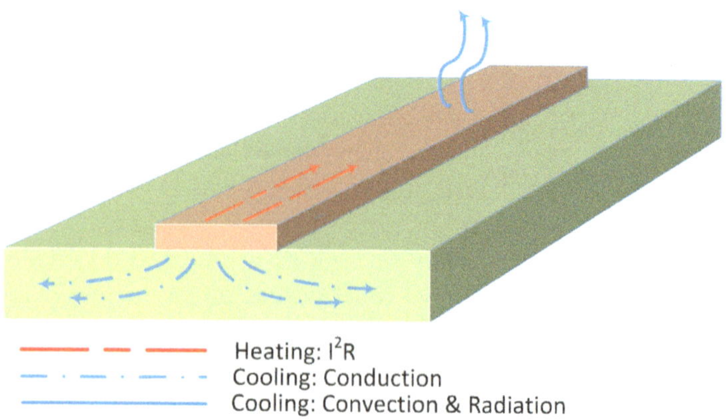

Figure 4.1 Trace heating and cooling dynamics.

100 watt-hours. We pay for *energy* usage in our homes (not power usage), typically based on a charge per kilowatt-hour (kWh).

4.3.2 Trace Heating

The I^2R power dissipation in a trace is a point concept. It reflects the amount of energy being dissipated at that *point* in time, and at that *point* along the trace. It is at that point in time because if voltage (for example) is changing, the power will also be changing. It is at that point along the trace because the trace parameters may change at any point along the trace. The total energy dissipated in the trace is the integral of the power dissipated in the trace times the time it is being dissipated, integrated over the length of the trace.

Resistance, as we covered in Chapter 3 (Equation 3.1), is resistivity times length and divided by cross-sectional area. If the area changes *at any point* along the trace, the resistance changes *at that point*, and the power dissipation *at that point* will also change. Similarly, if resistivity changes *at any point* along the trace (perhaps because of contamination of the copper), then the power dissipation *at that point* will also change. Thus, a trace does not necessarily heat uniformly along the trace. There may (and probably will) be some places hotter than other places along the trace, particularly at higher temperatures (see more about this in Chapter 13).

4.4 Trace Cooling

Traces on the top and bottom layers of a board have roughly half their surface area exposed to the air, and half in contact with the board material.[2] These traces cool by conduction, convection, and radiation. Internal traces are surrounded by board material and cool almost exclusively by conduction.

The ability of the board material to conduct heat away from the trace is largely determined by the thermal conductivity of the material. This conductivity can proceed in two directions: (a) away from the trace in the *x,y* plane (parallel to the trace and to the surface of the board) and in the *z*-axis (perpendicular to the trace and to the surface). These directions are often referred to as in-plane and through plane, respectively. Different board materials will have different thermal conductivity coefficients. These coefficients may or may not be included in the material specifications from the material supplier. If the coefficients are not available, they may be estimated, but thermal modeling becomes more difficult.

Thermal conductivity coefficients of dielectrics are typically in the range of 0.3 to 0.8 W/mK (watts per meter per degree K). Some newer materials, impregnated with ceramics, may have much higher thermal conductivities. Higher thermal conductivity means the board conducts heat away from the trace better.

The transfer of heat away from the board into the air is determined by the heat transfer coefficient (HTC).[3] HTC can be a little tricky to estimate. First, it is a function of the temperature difference between the trace and the air (e.g., it would increase if the air were cooled, and might decrease as the air around the trace heats up). Thus HTC is not necessarily constant during an investigation as the trace temperature increases. Second, it is a function of the velocity of the air (i.e., if a breeze is blowing). Thus, HTC increases if the trace is cooled by a fan. In any experimental investigation it is important to take measurements in still air and to wait long enough for the temperature differences between the trace and the surrounding air to stabilize.

The heat transfer characteristics of convection (through the air) and radiation (into space) are approximately equal under laboratory conditions. Thus, we would expect surface traces operated in a vacuum to cool about 50% less effectively than they would in a nonvacuum condition. That means we would expect traces in a vacuum would be hotter than traces not in a vacuum. And in fact we see that in the IPC data. We will see in Chapter 6 that that is confirmed in simulation models.

It turns out that the thermal conductivity of board materials is better than the heat transfer coefficient into the air. Therefore, internal traces (surrounded by board material) cool *better* than traces on the surface. This was pretty much unrecognized until IPC-2152 was published in 2009. Before that, the standard convention was to derate internal traces by 50% on the assumption that they cooled much less effectively than traces cooled by air.

4.4.1 Conductive Cooling

Our hypothesis is that the fiberglass in dielectrics conducts heat better than does the resin. Therefore, a board material with layers of fiberglass embedded in resin tends to cool better in the horizontal (or in-plane, or *x,y* plane) direction. And the cooling flow tends to follow the fiberglass layers (Figure 4.2). As a result, board materials are anisotropic; that is, they conduct heat better in one direction than in another. (See Appendix A for more on thermal conductivity and how it is measured.)

Finally, narrower traces cool (relatively speaking) better than wider traces. The reason is illustrated in Figure 4.3. During the cooling process, heat must conduct, or flow, out from under a trace. As shown in the figure, the transfer path is significantly longer for a wider trace than it is for a narrower one.

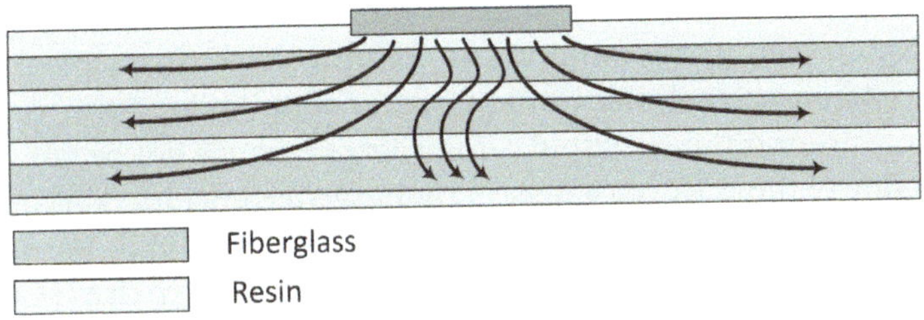

Figure 4.2 Cooling tends to follow the fiberglass layers in dielectrics.

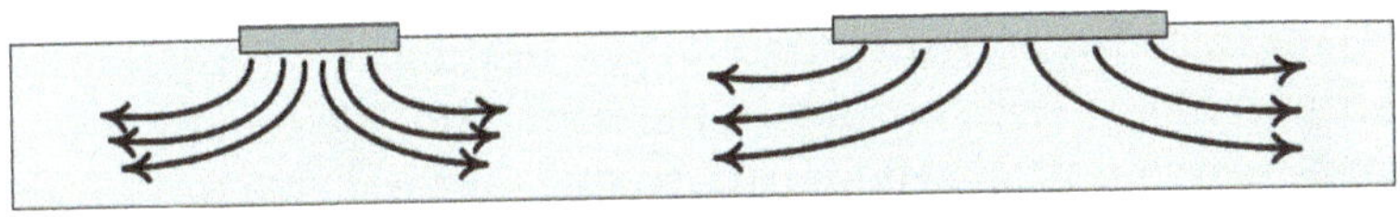

Figure 4.3 Narrow traces have a shorter cooling path compared to wider traces.

This is also evident in simulation results. Figure 4.4 illustrates the thermal plume around a trace. Both traces in the figure are carrying enough current to heat them to about 70°C. The figure illustrates how the temperature

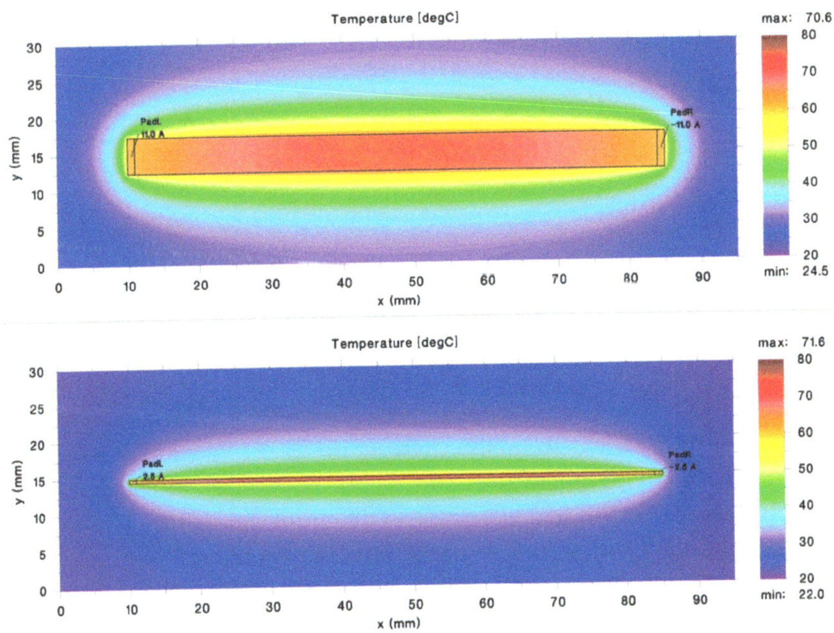

Figure 4.4 Thermal plumes around a 200 mil wide trace (top) and 20 mil wide trace (bottom).

gradient spreads out on the surface of the board away from the trace. In these models, the plume around the 200-mil (5-mm) trace extends out roughly 10 mm on each side of the trace, or about 4 times the trace width. The plume around the 20-mil (0.5-mm) trace extends out roughly 7.5 mm on each side of the trace, or about 30 times the trace width. This illustration applies to a model where the temperatures have stabilized. We will see in Chapter 12 that in situations where the trace temperature rises very quickly (as in a fusing situation) the difference is even more dramatic.

4.5 Mathematical Model of Trace Heating and Cooling

A constant trace temperature occurs when heating equals cooling. Heating is a function of power dissipation in the trace, I^2R. The cooling relationship is complicated, but is related, in part, to trace surface area or to width plus thickness ($W + Th$). So, the change in temperature will be proportional to I^2R and inversely proportional to ($W + Th$), or

$$\Delta T \propto \frac{I^2 R}{(W + Th)} \tag{4.3}$$

where

ΔT = change in temperature

I = current

R = trace resistance

Th = trace thickness

W = trace width

4.6 Role of Current Density

There are some who believe that trace temperatures can be correlated to current density (A/in^2). Actually, there is no direct relationship between current density and temperature. This will be covered in much more detail in Chapter 14. Here are two examples to illustrate the point.

Case 1: Assume two traces, A and B, are each carrying the same current. They each have the same cross-sectional area. Therefore, the current densities are identical. But trace A is wide and thin and trace B is narrow and thicker. Trace A will be cooler because it has more surface area from which to dissipate heat.

Case 2: Assume two traces, A and B, are each carrying the same current and have identical cross-sectional areas and form factors. The current density will be identical in each trace. The resistivity of the conductor material making up trace A is higher than the resistivity of the conductor material making up trace B. Therefore, the resistance of trace A will be higher than trace B. Therefore, the I^2R dissipation in trace A will be greater than in trace B and trace A will be hotter.

Therefore, there is no necessary relationship between current density and temperature.

4.7 Measuring Trace Temperature

There are several different ways we can determine the temperature of a trace. None is without problems, but each has some individual strengths. This section will outline a few of these.

4.7.1 IPC Procedure

One way to measure the trace temperature is to calculate the resistance of the trace at an elevated temperature and compare it to the resistance at a reference temperature (the change in resistance method). The IPC outlines this test procedure in its procedure manual.[4]

A standard test trace is defined as shown in Figure 4.5.

The trace length from end to end is 12 inches. The pads at each end are 200 mil in diameter. Three inches in from each end (dimension A) is a sense pad, 20 mil wide by 400 mil long. That leaves the active test length (dimension B) at 6 inches. Sense leads, #26 AWG magnet wire, are soldered to the sense pads. Then the procedure follows these steps:

1. Place the board horizontally in still air.
2. Measure the resistance of the trace at the reference temperature (ambient) by passing a known and constant current through the

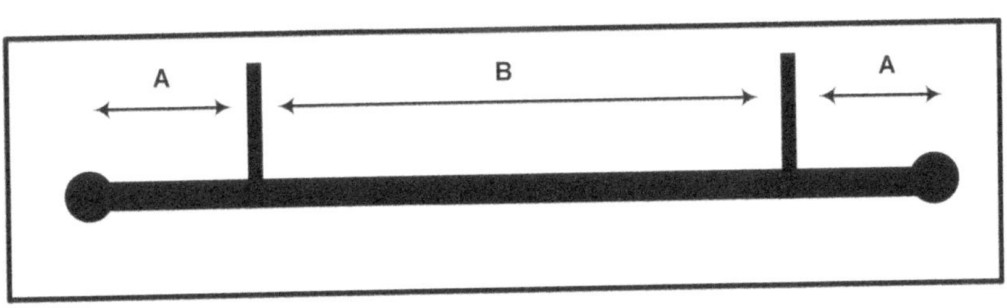

Figure 4.5 IPC standard test trace (not to scale).

trace no larger than 100 ma. Use Ohm's law (R = v/i) to calculate the reference resistance. (Alternatively, the resistance could be measured with a precision ohmmeter that is electrically isolated from the trace).

3. Apply a known (and constant) current to the trace under test.

4. Measure the voltage across the sensing points. Calculate the resistance at the elevated temperature, again by Ohm's law (or again use a precision ohmmeter).

5. Calculate the change in temperature using (4.4) (refer back to Section 3.5):

$$\Delta T = \frac{1}{\alpha_0}\left(\frac{R_t}{R_{t_0}} - 1\right) \tag{4.4}$$

where

ΔT = change in temperature

t_0 = reference temperature

R_t = resistance at the temperature of interest

R_{t0} = resistance at the reference temperature

α_0 = thermal coefficient of resistivity at the reference temperature

This procedure measures, by definition, the *average* temperature along the trace. We will show later that there is virtually always a gradient in temperature along a trace, especially near the ends (pads) On the other hand, the fact that the sense leads are spaced 3 inches in from each end is intended to minimize the effects of any gradients in the IPC data. Therefore, it is important to note that the IPC procedure measures the *average* temperature accurately, but does not account for gradients or hot spots.

4.7.2 Infrared Measurement

Another approach for measuring trace temperature is to use an infrared, or heat-sensing, microscope or camera to measure the temperature at a point along the surface of the trace. The strength of this approach is that it is noncontact and it measures the temperature at a *point* along the trace, which is often what we really want. Later in this book, especially in Chapter 13, we will show numerous examples of trace temperature measurements using an infrared (thermal) camera.

But there are also difficulties with this approach. For one, it assumes that the emissivity of the trace surface is known, which is not always true. Second, it is not as precise as other measures, often with uncertainties measured in several degrees C. On the other hand, this may be the only viable approach for measuring trace temperatures that are very high (as will be covered in Chapter 12). Modern heat-sensing measurement tools are getting better and better at overcoming some of the inherent problems with this approach.

4.7.3 Thermocouple Measurement

A third approach to trace temperature measurement is the use of a thermocouple. This is, by definition, a point measurement approach. A thermocouple is placed directly on the trace and the temperature measured directly (Figure 4.6).

This has several advantages. From a data measurement standpoint it is relatively quick, and from a location standpoint it is very precise. There are data loggers readily available that can record temperatures as they are measured and automatically graph them. Thermocouples are available with probe points as small as 3 mils and with response times as short as a second

Figure 4.6 Use of a thermocouple to measure trace temperature.

or two. However, there is the possibility that the contact with the trace may impact the measurement. One view is that the probe may conduct heat away from the trace and lower the reading. On the other hand, the thermocouple has so small a thermal load, and the copper conductivity of the trace is so high, that any loading may be minimal.

Care must be taken that the thermocouple makes good, solid contact with the trace. And the temperature range of many thermocouple probes (because of their coatings) may top out at only around 300°C to 350°C.

4.7.4 Point versus Average Measurements

The change-of-resistance approach to temperature measurement (the IPC approach) measures the average temperature of the trace (Figure 4.7). Typically, there is a thermal gradient along a trace; traces are coolest at the pads at each end of the trace and hottest at the center (see Chapter 7). The average temperature might not be what we want. We might want to know what the maximum temperature is along the trace. The IPC test procedure tries to mitigate this by measuring the temperature along the 6-inch center portion of the subject trace.

Infrared and thermocouple measurements record the temperature at a point along the trace. But since traces do not heat uniformly (see Chapter 13), the selected point might not be the hottest point along the trace. Therefore, the most suitable measurement technique is not always obvious.

4.8 Trace Temperature Curves

When current is applied to a trace, the temperature increases. But there is a time constant associated with that increase. The pattern of temperature versus time typically takes one of three general shapes depending on whether the trace is severely overloaded.

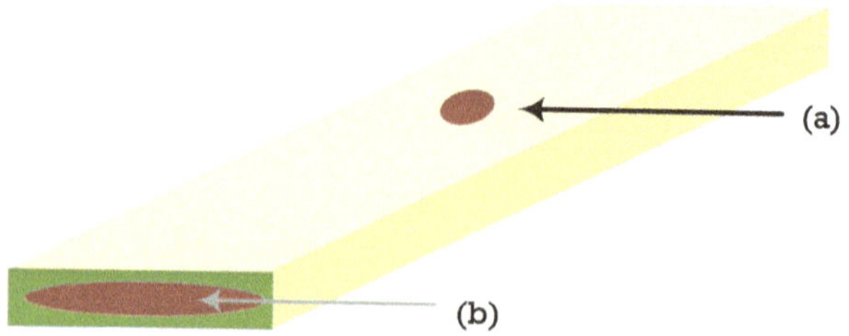

Figure 4.7 Point (a) vs. average (b) temperature measurements.

4.8.1 Typical Curve

When current is applied to a trace, the temperature of the trace rises to a point and then stabilizes (Figure 4.8). The time for stabilization (A) is typically anywhere from about 6 minutes to 15 minutes. After that, the temperature remains stable.

4.8.2 Heavy Overload

A trace subject to a heavy current overload (so that it melts fairly quickly) follows a pattern typically like that shown in Figure 4.9. The temperature rises in a (more or less) linear way until it melts at some point. This is typical of traces that reach a melting temperature within a few (say less than 5 or 10) seconds.

4.8.3 Marginal Overload

Let's say we have a trace that is carrying so much current that it needs supplemental cooling. Cooling may be provided by heat sinks or by forced air. Then let's say that cooling degrades for some reason. Or, let's say the trace is carrying just enough current to eventually fuse. This type of failure tends to follow a pattern similar to that shown in Figure 4.10.[5]

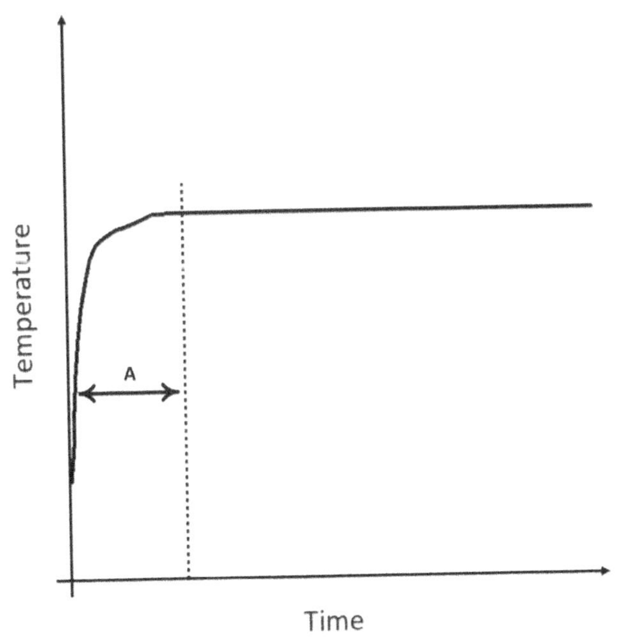

Figure 4.8 Typical temperature relationship.

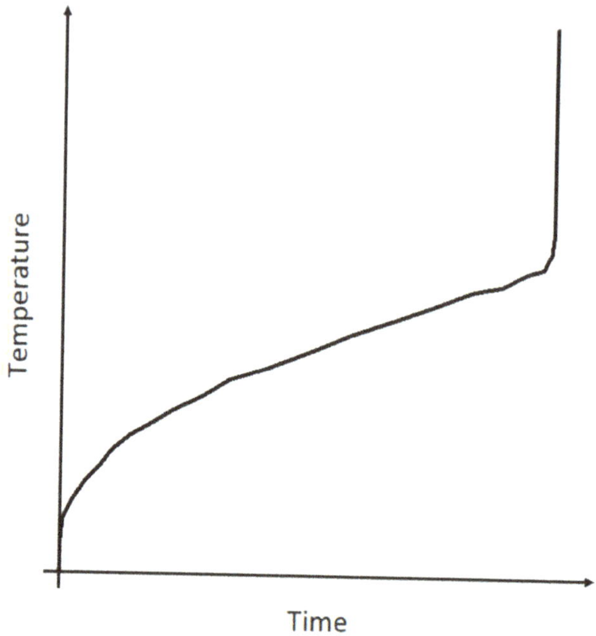

Figure 4.9 Typical fusing curve.

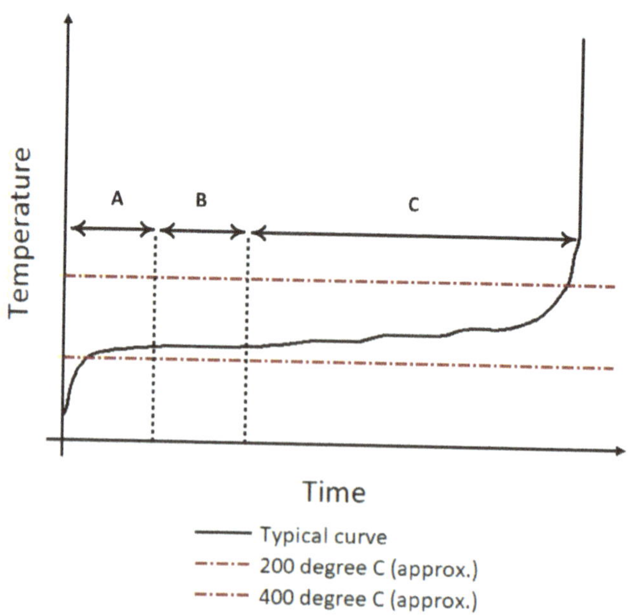

Figure 4.10 Long-term trace failure.

Over the time frame shown as "A" in the figure, the temperature rises to a point and stabilizes. This is exactly like Figure 4.8 and happens over the same time frame. Then the temperature seems stable over time frame "B," which can last from minutes to hours. But then the temperature begins to rise. The reason may be the result of several things. Maybe the support cooling approaches began to degrade. Maybe the dielectric began to change (see Section 2.4). But for whatever reason, the temperature begins to increase until the trace ultimately fails (melts) at some spot. This is sometimes (inaccurately) described as thermal runaway.

The temperature where time range C might start is very situation-specific and will vary from trace to trace. But the range typically is around 220ºC to 270ºC. Once runaway begins, failure is inevitable (unless the current is lowered). Time frame "C" is typically around 15 minutes or so, but it is very difficult to generalize. We will cover this topic in much more detail in Chapter 12.

End Notes

1. Thermodynamic engineers make a distinction between the initial temperature (initial condition) and ambient temperature since the initial condition might, for some reason, be higher or lower than ambient. Electrical engineers are more casual about this and usually (perhaps incorrectly) equate the initial condition with ambient temperature.

2. To the extent that these traces are coated (say conformal or solder coated for example), the cooling may be enhanced or restricted somewhat.

3. See https://en.wikipedia.org/wiki/Heat_transfer_coefficient for an explanation of HTC.

4. See *IPC-TM-650 Test Methods Manual*, Number 2.5.4.1a, "Conductor Temperature Rise Due to Current Changes in Conductors," freely downloadable at http://www.ipc.org/test-methods.aspx.

5. There is a curve very similar to this in Codreanu, N., R. Bunea, and P. Svasta, "New Methods of Testing PCB Traces Capacity and Fusing," Politehnica University of Bucharest, Center for Technological Electronics and Interconnection Techniques, UPB-CETTI, internal research project and report of UPB-CETTI/Winter-2010, (Figure 6), http://www.cetti.ro/cadence/articles/New_Methods_of_Testing_PCB_Traces_Capacity_and_Fusing.pdf.

CHAPTER

5

Contents

IPC Curves

5.1 Bottom Line

- The IPC-2152 curves represent the best experimental set of data ever assembled on PCB trace current/temperature relationships.
- In this chapter we look at what the curves tell us and fit the curves with equations.

5.2 IPC-2152

IPC 2152, "Standard for Determining Current Carrying Capacity in Printed Board Design," provides the best data on PCB trace current/temperature relationships that have ever been assembled. The publication has over 90 pages and over 75 charts and tables. It covers external traces, internal traces, and traces in a vacuum, as well as traces from 0.5-oz to 3.0-oz thickness.[1]

5.3 Measuring the Temperature

The IPC outlines a test procedure that it follows in this type of investigation.[2] A standard test trace is defined as shown in Figure 4.5 and reproduced here as Figure 5.1.

The trace length from end to end is 12 inches. The pads at each end are 200 mil in diameter. Three inches in from each end (dimension A) is a sense pad, 20 mil wide by 400 mil long. That leaves the active test length (dimension B) at 6 inches. Sense leads, #26 AWG magnet wire, are soldered to the sense pads.

Although the details in the procedure may seem tedious, they boil down to the following steps:

1. Place the board horizontally in still air.
2. Measure the resistance of the trace across the sensing points at the reference temperature (ambient) by passing a known and constant current through the trace no larger than 100 ma. Use Ohm's law ($R_{to} = v/i$) to calculate the reference resistance.
3. Determine the cross-sectional area of the trace using known trace dimensions or by using (5.1).

$$A = \rho * L / R_{to} \tag{5.1}$$

where

A = cross-sectional area

ρ = resistivity of the conducting material

L = length of the trace

R_{to} = the measured resistance

4. Apply a known (and constant) current to the trace under test.
5. Measure the voltage across the sensing points. Calculate the resistance (R_t) at the elevated temperature, again by Ohm's law.

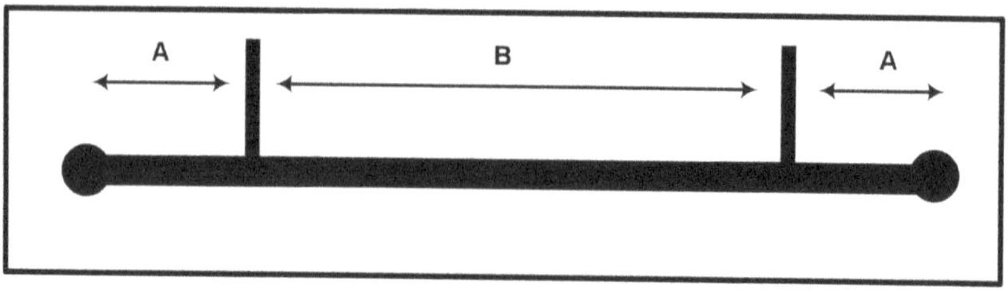

Figure 5.1 IPC standard test trace (not to scale).

6. Calculate the change in temperature using (5.2) (refer back to Section 3.3).

$$\Delta T = \frac{1}{\alpha_0}\left[\frac{R_t}{R_{t_0}} - 1\right] \qquad (5.2)$$

where

ΔT = change in temperature

t_o = reference temperature

R_t = Resistance at the temperature of interest

R_{t0} = Resistance at the reference temperature

α_0 = Thermal coefficient of resistivity at the reference temperature.

7. Repeat step 6 for a variety of currents and voltages and record the data.

There are various strengths and weaknesses with this approach. For one, the IPC approach seems to suggest that ρ, the resistivity, can be taken as published. It does not suggest that it be verified. It also suggests that α_o be calculated if unknown but it also seems to suggest that α_o can be determined from published values. As we saw in Chapter 3, both of these values can be suspect and subject to error. (But, to be fair, this comment applies to any trace, anywhere!)

Finally, the procedure requires that the reference resistance (R_{to}) be measured with a digital multimeter at a current not to exceed 100 ma. For a larger trace (say a 2.0 oz, 100 mil wide trace that is 6″ long), this results in an expected resistance of perhaps 0.016 ohms, resulting in a measured voltage of only 1.6 millivolt (mV). A voltage that low is tough to measure with precision with a multimeter.

The IPC charts in IPC-2152 plot current versus cross-sectional area for constant temperature curves. But the data that is taken is for current and temperature for constant cross-sectional area. Thus, data aggregation and interpolation steps (across various test samples) are required before publication. (Again, to be fair, this is a practical and acceptable procedure. It is noted simply to point out that there is an opportunity for error to be introduced in these several steps.)

This procedure measures, by definition, the *average* temperature along the trace. We will show later that there is virtually always a gradient in temperature along a trace, especially near the ends (pads). On the other hand, the fact that the sense leads are spaced 3 inches in from each end is an attempt to minimize the effects of any gradients in the IPC data. So the IPC procedure measures the average temperature accurately, but does not account for gradients or peak temperatures.

5.4 IPC Curves

There are three sets of curves in IPC-2152, for external, internal, and vacuum environments. Although similar, there are important differences between each of them. We will look at them separately.

5.4.1 External Results

The results from the testing are provided in curves like that shown in Figure 5.2. The curves plot current as a function of cross-sectional area and create constant-temperature (change) lines on the graph. Each set of curves represent a constant trace thickness.

It will be useful to modify these curves so that they plot the change in temperature as a function of current and have constant-width curves on the graph. That is, in part, because it is more useful for PCB designers to have data presented this way. It is also because our model (see (4.3)) is expressed this way. Finally, it is also because thermal simulations tend to

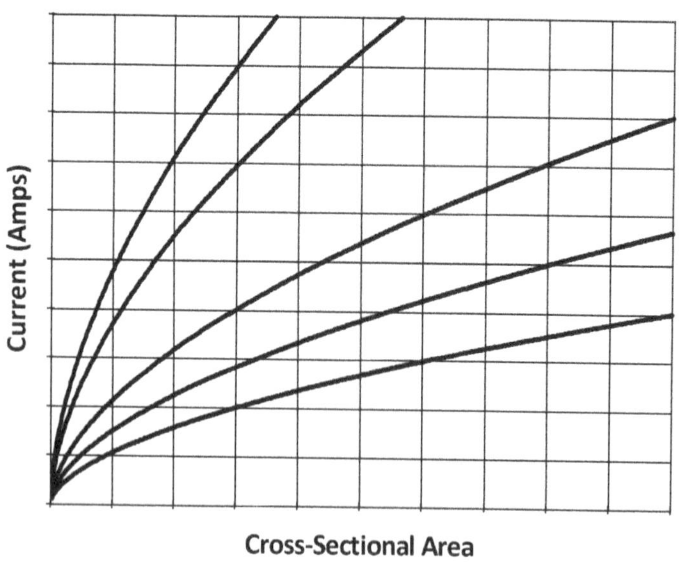

Figure 5.2 Typical form of IPC curves. Lines are constant temperature.

be formulated this way. So we need to transpose the IPC data to a different format. When we do that, we obtain a set of graphs that look like Figure 5.3. (All the curves in this set have the same thickness.)

5.4.2 External IPC Data Equations

The next task is to try to fit these curves with an equation. One way to do this is with a multiple regression analysis using a resource such as any current spreadsheet. Another way is to start with our model ((4.3), repeated here as (5.3)):

$$\Delta T \propto \frac{I^2 R}{(w + Th)} \tag{5.3}$$

Recognize that resistance, *R*, is inversely proportional to area (width * thickness), so this model can also be expressed as (5.4):

$$\Delta T \propto \frac{C^2}{Width^{a1} * Th^{a2}} \tag{5.4}$$

where

ΔT = change in temperature

C = current

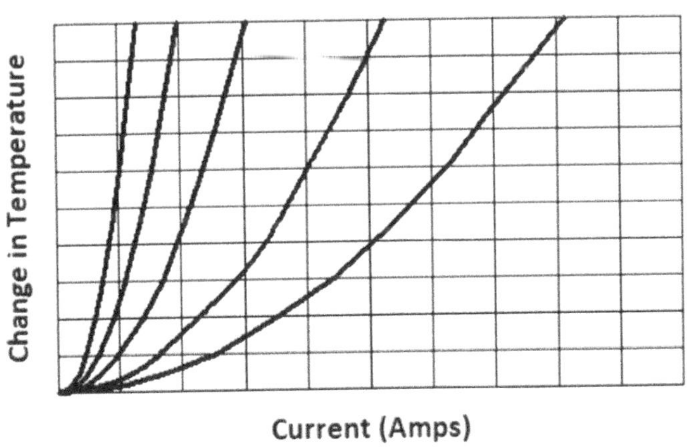

Figure 5.3 IPC curves redrawn to a different set of axes. Lines are constant width.

Width = trace width

Th = trace thickness

*a*1 and *a*2 are undetermined constants that are approximately 1.0 to 1.5

After a little work it was determined that the best fit for these curves was an equation of the form

$$\Delta T = 215.3 * C^2 * W^{-1.15} * Th^{-1.0} \tag{5.5}$$

Plotting this equation onto a figure like Figure 5.3 (with the appropriate values for *W* and *Th*) gives the results shown in Figure 5.4 (the solid lines are derived from the IPC data and the dotted lines are the equations).

This result is impressive, but what would be even more meaningful would be if this same equation, for 2.0-oz external data, would also fit the 3.0-oz external IPC data. If that were the case, we would have a fair amount of confidence in both the IPC data and the equations.

Figure 5.5 illustrates similar results for the 3.0-oz external IPC data. The curves were generated in the same fashion as the 2.0-oz curves, and the equation is the same, differing only in terms of the appropriate variables (width and thickness).

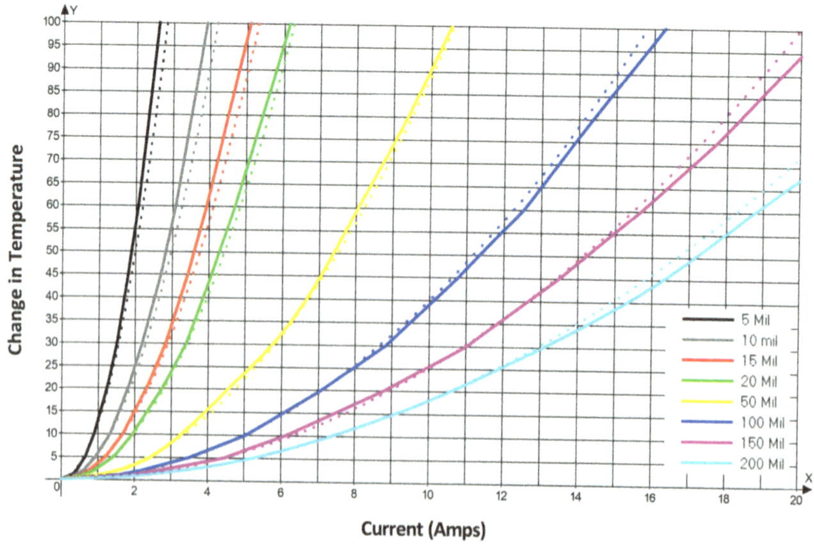

Figure 5.4 2.0-oz external curves (solid lines) fitted with (5.5) (dotted lines).

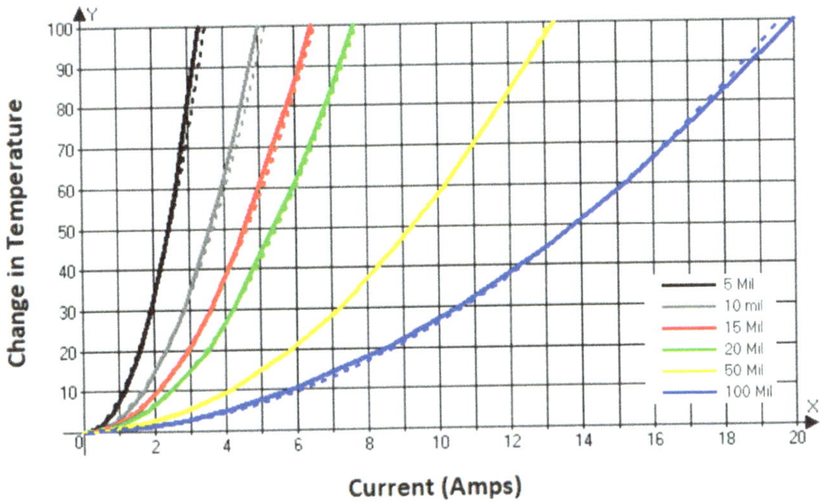

Figure 5.5 3.0-oz external curves (solid lines) fitted with (5.5) (dotted lines).

The fits are obviously extremely good. This gives us very high confidence in the equations.

We can't do a similar fit test for 1-oz traces because there is no separate 1-oz external data in IPC 2152. Section 5 in IPC 2152 provides a set of charts for overall use, but they reflect a consolidation of all data, not exclusively 1-oz data. Similarly, there are no 1-oz external data charts in the IPC 2152 appendices. We have produced Figure 5.6; however, based on (5.5). In the next chapter we will fit this curve with thermal simulation results for comparison with the 2-oz and 3-oz curves.

As an aside, it is worth noting that (5.5) also proves that the change in temperature is not a function of current density alone, as some people might think. Since the coefficients for the width and thickness terms are different, the form factor of the trace (i.e., whether it is wider or thicker for the same cross-sectional area) matters. Although the coefficients are not exactly what we expect from our initial model (for reasons that are not exactly clear), the closeness of the fit (and the fit of the thermal models in the next section) cannot be ignored.

5.4.3 Internal IPC Data Equations

A real surprise from the IPC 2512 study and data was that the internal races actually cool *better than* the external traces do. The internal traces are a little cooler than the external traces.

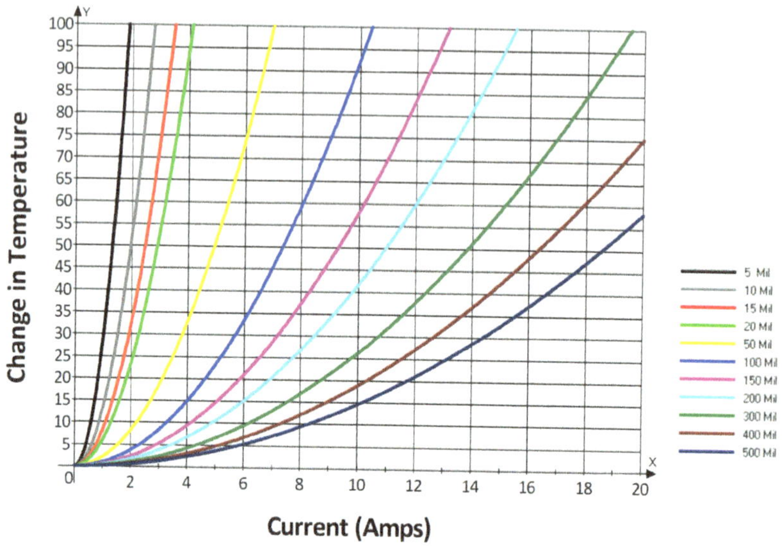

Figure 5.6 1-oz curves generated from (5.5).

Fitting the internal IPC data does not go quite so smoothly. To save space in this book we have put the four internal fitted curves in Appendix C. On the surface, the curves and the fits look excellent. The problem is, in this case, the equations are not all exactly equal. They are quite close, but not exactly equal. The primary difference is in the constant term. Table 5.1 provides the coefficients for the equations.

We will look at how the thermal models fit the internal data later on. At this point, our suspicion is that the differences reflect a control issue in the study. As we will see in Chapters 6 and 7 the temperature of a trace is very sensitive to a variety of factors, only some of which are easily controlled. Two of those are trace thickness and dielectric thermal conductivity coefficients. If these parameters were not well-controlled, variations in the fit equations might well occur. The differences here between trace widths are not nearly as great, however, as some of the other differences we will explore in Chapter 6.

5.4.4 IPC Vacuum Data

Traces are cooled by three effects: (1) heat conducting away from the trace through the board, (2) heat convected away from the board into the surrounding air, and (3) heat radiating away from the board into space. In a vacuum, there is no surrounding air, so heat that gets to the surface (either because the trace is on the surface or because the heat has conducted through the board to the surface) can only radiate away. This is much less efficient than convection through the air, so traces in a vacuum understandably run quite a bit hotter than otherwise. And it appears not to make too much difference whether we are talking about internal or external traces. (IPC 2152 does not provide separate data for internal and external traces in a vacuum or provide any 1-oz data for traces in a vacuum.)

The curves for traces in a vacuum (and for their fitted equations) are provided in Appendix C. Again, the fitted equations are not exactly equal for all the widths. And again, control issues might be part of the explanation for this.

Figure 5.7 illustrates the difference between selected 2-oz external, internal, and vacuum traces. The trace in a vacuum is always hotter than the same trace in external air. The internal trace is always cooler than the same trace in external air. The pattern is similar for all other trace thicknesses.

Table 5.1 shows the approximate equations used for fitting the curves. The complete set of equations is provided in Appendix D.

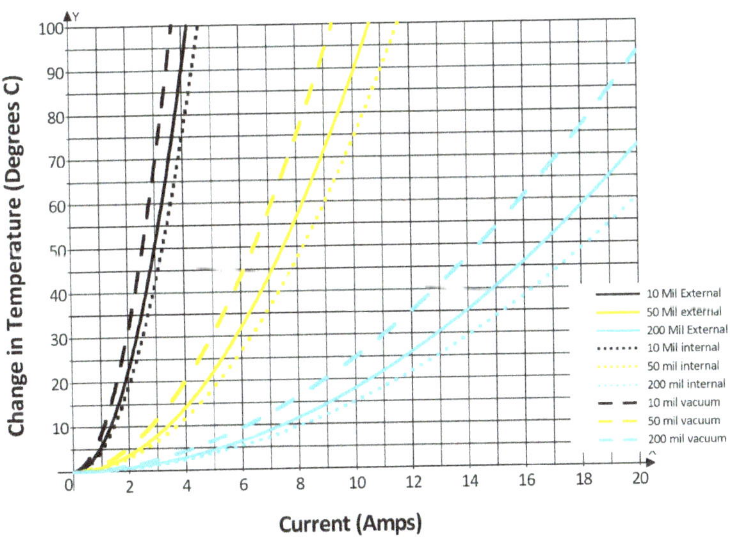

Figure 5.7 Comparison of 2-oz external, internal, and vacuum traces of width 10 mil, 50 mil, and 200 mil, respectively.

Table 5.1
Coefficients for all the IPC Equations

Data	Constant	C^	W^	Th^
External				
All	215.3	2	−1.15	−1.0
Internal				
0.5 oz	110–130	2	−1.10	−1.52
1 oz	200	1.9	−1.10	−1.52
2 oz.	300	2	−1.15	−1.52
3 oz	429–368	1.9	−1.10	−1.52
Vacuum				
0.5 oz	210–235	1.9	−1.10	−1.52
2 oz	480	1.9	−1.10	−1.52
3 oz	460	1.95	−1.15	−1.52

Copper trace thickness is typically measured in ounces of copper. A 1.0-oz trace is the thickness of an ounce of copper spread out over a 1 square foot area. A 1-oz trace is approximately 1.3 mils thick.

End Notes

1. IPC 2152, "Standard for Determining Current Carrying Capacity in Printed Board Design," 2009, IPC.com.
2. See *IPC-TM-650 Test Methods Manual,* Number 2.5.4.1a, "Conductor Temperature Rise Due to Current Changes in Conductors," freely downloadable at http://www.ipc.org/test-methods.aspx.

CHAPTER

6

Contents

Thermal Simulations

6.1 Bottom Line

- Thermal simulation models can accurately simulate the curves in IPC-2152. This gives us confidence that they can also simulate real-world trace configurations.

6.2 Background

In the previous chapter we looked at the IPC curves and fit them with equations. In this chapter, we will independently try to verify the curves using a computer thermal simulation program. If we are successful at doing that, then in Chapter 7 we will simulate other typical conditions not covered by the IPC curves—things like adjacent trace and underlying planes—and see what impact those things might have on the results.

6.3 Modeling Traces

Dr. Johannes Adam has written ADAM Research's Thermal Risk Management (TRM) simulation software for thermally characterizing printed circuit boards. TRM was originally conceived and designed to analyze temperatures across a circuit board, taking into consideration the complete trace layout with optional joule heating as well as various components and their own contributions to heat generation. Although the program could be adapted to the measurement of an individual trace, it was not originally conceived with that use in mind. Consequently, a couple of adjustments and adaptations had to be made for our particular analyses.

For those of us in the printed circuit board industry, a thermal simulation model is to trace temperature calculations what a field effect model is to trace impedance calculations (or what a meteorological model is to weather forecasting). And the approach is not too different. The 3-D structure is analyzed by first looking at a tiny cube[1] within the structure (see Figure 6.1). We assume the cube is small enough that the conditions inside the cube are uniform (homogeneous). At this micro level, the computations for the boundary conditions between the cubes are relatively straightforward to define. So you do the calculation for one cube, then the next cube, and then the next one, and so on. And you keep going in this iterative process until you have solved for the entire structure.

Figure 6.1 represents a very small portion of the model we will develop in Section 6.2. There will be a very large number of cubes in this analysis, and a great many calculations for each. The analysis becomes a very large matrix algebra computational problem; very difficult for an individual but

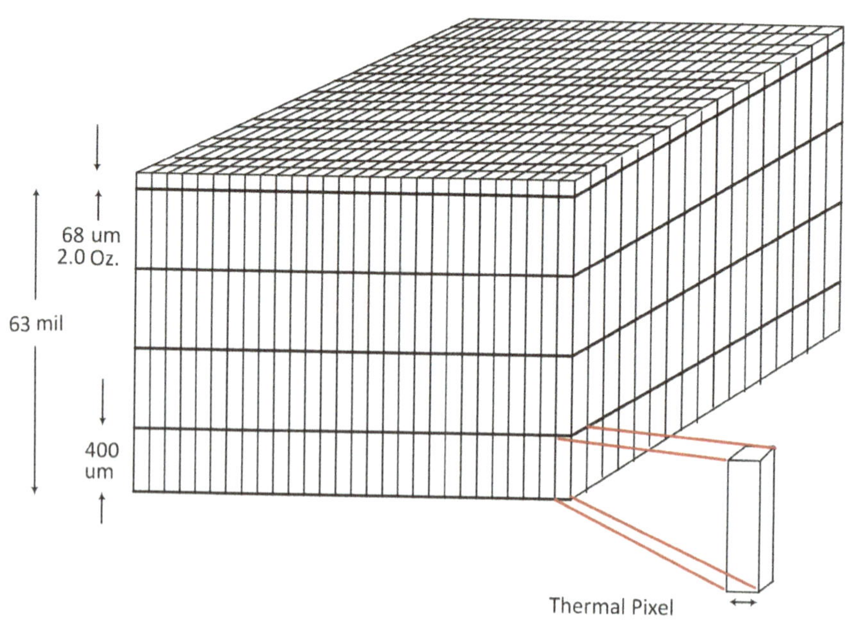

Figure 6.1 A model consists of a large number of square or rectangular cubes.

perfect for a computer. Any practical problem can require a large number of resources. A reasonable problem, with reasonable accuracy, can require a few minutes to several hours to solve, even with a powerful 64-bit desktop computer.

We will illustrate a couple of the adjustments we had to make below as we describe one of the models. But two comments deserve mention up front.

First, as mentioned in Section 3.3, resistance increases with temperature. So if we initiate an analysis with a certain resistivity in mind, and reach an elevated temperature, the resistance will have changed at that elevated temperature. That means the trace will be hotter than the analysis calculates. TRM runs in two selectable modes. One goes directly to a solution and one takes changing resistivity into account.

In the second mode, the program executes with initial values and reaches a solution. Then the program recalculates the resistance of the trace at the new temperature and executes (loops) a second time. And it keeps doing this. The user sets the number of loops as part of the setup. Experimentally we determined that two to three loops were sufficient to reach (acceptable) stability at lower temperatures and for internal traces and four loops (or more) were needed at higher temperatures.

Second, setting up a model requires defining the HTC, which was introduced in Chapter 4. This is a coefficient that defines how effectively a surface transfers heat to another medium. In this situation, it refers to how effectively the board and trace transfer heat to the surrounding air. This coefficient, found in the literature with letter h, is often not familiar to electrical engineers (other than perhaps the thermal resistance, Rth). However, h only provides the necessary coupling between temperature inside the board and the ambient (otherwise heat could not leave a board). In practice HTC contains the contributions of convection *plus* radiation. While HTC can be estimated quite well for flat and uniformly heated plates, it is not always intuitive what the value should be for a single trace on a board. But more importantly, HTC increases with temperature, and it is not at all clear how it increases for an individual trace. Here is how we attacked this problem.

We set up a model and entered a reasonable value for HTC. We solved the model and compared the result to the equations established in Chapter 5. We found the base value (i.e., the value of HTC for low currents and temperatures) pretty quickly. It was[2] 10 W/mK. Then we solved the models at higher currents and determined what values of HTC were required to fit the equations. Higher values were expected and determined.

At lower currents (temperatures) the model results were pretty insensitive to HTC, often differing by 1 degree C or less. At higher currents (temperatures) a range of HTC from 11 to 14 only resulted in a temperature difference of around 10 degrees (in one case 20 degrees). So the determination of the correct value for HTC was more like tweaking the results of the model.

The range of HTC's needed for external and internal traces is shown in Figure 6.2. The range falls between the red lines.

While it may seem like we are somehow forcing the model to fit the data, think of the initial steps as more like *calibrating* the model. Once we knew what the predictable range was, we could then preset the HTC value and get acceptable results.

6.4 The Modeling Process

It is not our intent in this section to teach someone how to run a TRM thermal simulation model. We will go through the steps to illustrate what is involved, but we won't cover every single step in the process. We want to do this so the reader can see that the model does, in fact, control for a lot of the variables we have introduced in previous chapters, including resistivity, thermal conductivity coefficients, and even radiation. Furthermore, this illustrates only one software's approach. There are other software programs out there that can perform similar simulations. We are using TRM as an illustration here because that is what we used, and that is what we know.

The IPC 2-oz, 200-mil trace is modeled as shown at the top of Figure 6.7 (an actual image from the model's output folder). The modeled trace is

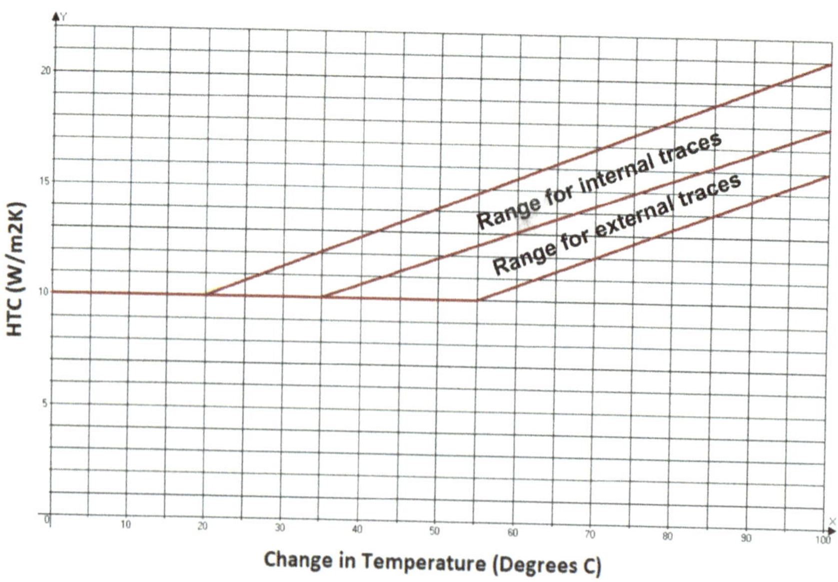

Figure 6.2 Approximate range of HTC for external and internal traces.

approximately 300 mm (11.8 in) long and 5 mm (200 mil) wide. The sensor pads are placed approximately 6.0 inches apart. There are 4.0-mm wide traces attached to the ends of the sensor lines to simulate the leads that are soldered onto the IPC test fixture.

Figure 6.3 shows the "Build Board/Prepare" menu for setting up the board. The board outline is 350 × 45 mm. The default values for the conductor and dielectric are specified (here normal copper and polyimide). The thermal pixel is specified as 0.2 mm. Think of the thermal pixel as the minimum length of the side of one of the cubes shown in Figure 6.1. This must be smaller than any other dimension in the *x,y* plane. The smaller the cube, the more cubes there are in the analysis and the larger are the computational matrices. A thermal pixel of 0.2 is not particularly small, and this analysis will run fairly quickly. If we were analyzing a via model, the thermal pixel might be specified as 0.02 (or maybe even smaller) and the analysis would take much longer.

Figure 6.4 shows a copy of the material database for polyimide. The arrows highlight the entries for the in-plane and through-plane entries for the thermal conductivity coefficient (these entries are editable).

Next is the "Expose and Laminate" menu seen in Figure 6.5. Think of this as the board stackup. Of note, the top layer (which will be copper foil) is 68 μm (2 oz). The board itself is 1,600 mm (63 mils) thick. It is broken into four separate layers of 400 mm each for technical reasons. (Figure 6.1

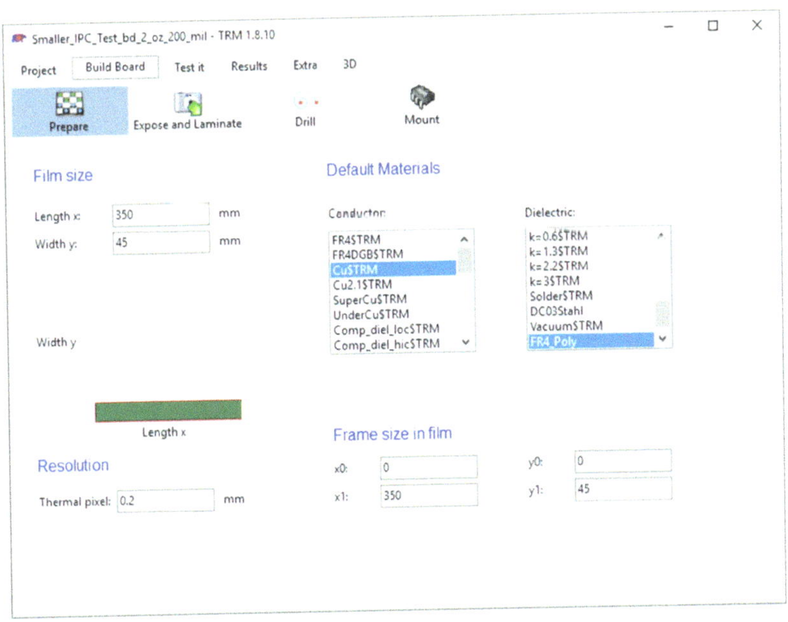

Figure 6.3 The project prepare menu.

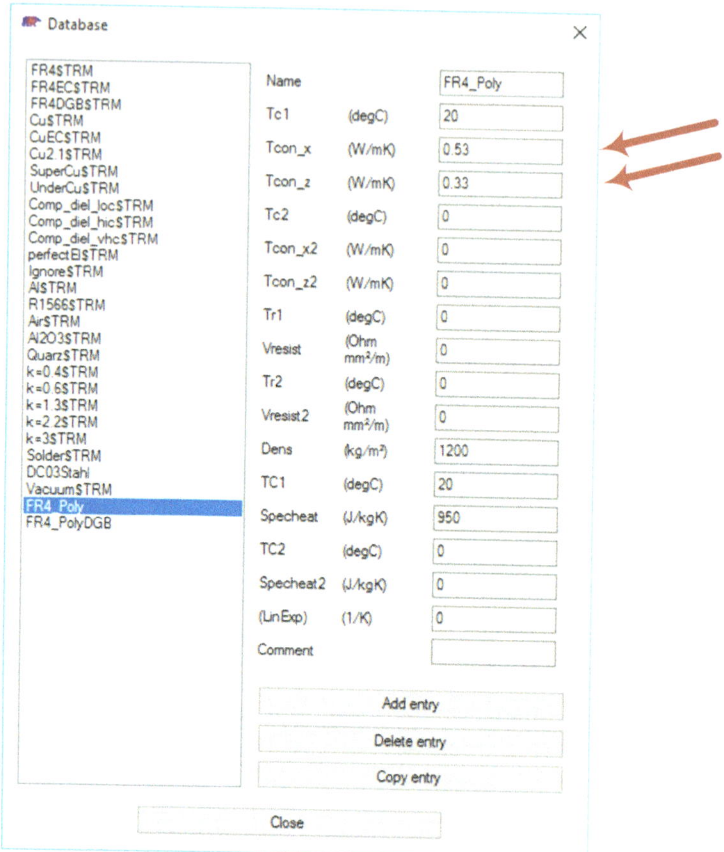

Figure 6.4 The materials properties database.

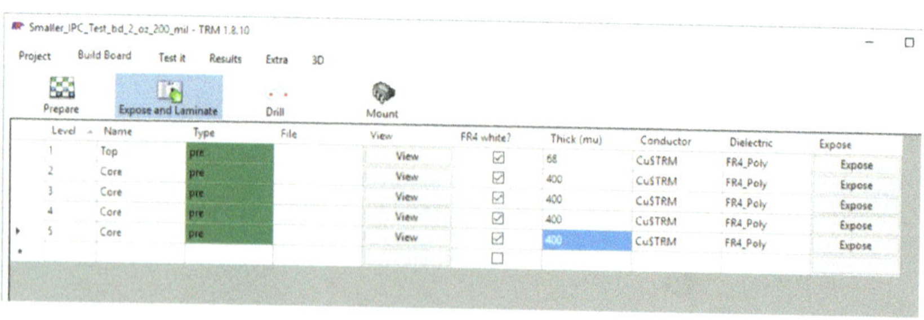

Figure 6.5 Expose and laminate menu.

models this stackup.) Adding multiple layers provides a little more precision to the model. The conductor and dielectric specifications set here override the default values set in Figure 6.3.

There are separate menus for specifying the conductor and dielectric values in very precise detail. Of importance to us would be the resistivity and thermal coefficient of resistivity values for the conductor, and the thermal conductivity values for the dielectric. Since we don't know the thermal conductivity of the specific type of polyimide used in the IPC studies, our model uses the default values for polyimide, which is 0.53 for both in-plane and through-plane conductivities.

Next is the Environment menu seen in Figure 6.6. This is pretty straightforward. The ambient temperature is set at 20°C and the HTC coefficient (discussed above) set as 11.

Then we come to the "Set Loads" menu seen in Figure 6.7. This is the heart of the setup. This is where we define the traces, pads, test points, and anything else we want to specify. Remember that the board is 350 mm wide and 45 mm high. Think of the lower, left-hand corner of the board as having coordinates, *x,y* equal to 0,0. The power and return pads start at coordinates 30,15 and 330,15, respectively. They are each 5 mm wide and 5 mm high. Sixteen amps enters the Power pad (+16) and exits the Return pad (−16). The trace starts at 33.2,15 and is 300 mm long and 5 mm high. Since the trace overlaps the pads, the trace length distance between the pads is 295 mm.

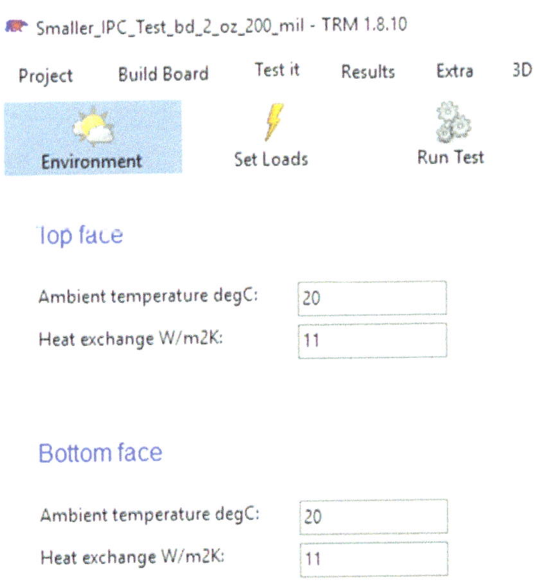

Figure 6.6 Environment menu.

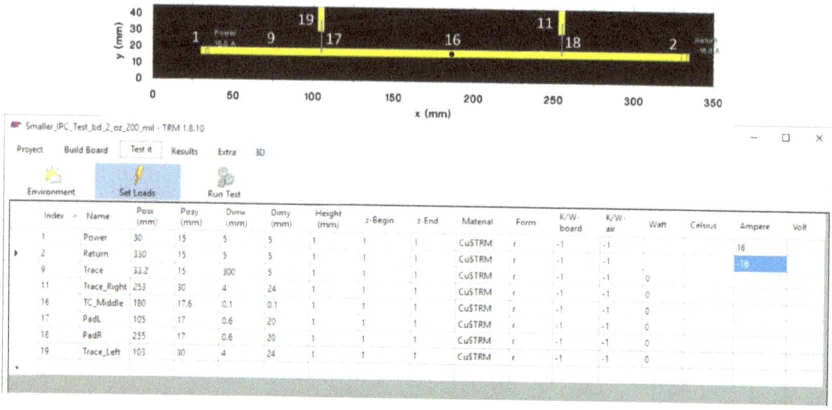

Index	Name	Posx (mm)	Posy (mm)	Dimx (mm)	Dimy (mm)	Height (mm)	z-Begin	z-End	Material	Form	K/W-board	K/W-air	Watt	Celsius	Ampere	Volt
1	Power	30	15	5	5	1	1	1	CuSTRM	r	-1	-1			16	
2	Return	330	15	5	5	1	1	1	CuSTRM	r	-1	-1			-16	
9	Trace	33.2	15	300	5	1	1	1	CuSTRM	r	-1	-1	0			
11	Trace_Right	253	30	4	24	1	1	1	CuSTRM	r	-1	-1	0			
16	TC_Middle	180	17.6	0.1	0.1	1	1	1	CuSTRM	r	-1	-1	0			
17	PadL	105	17	0.6	20	1	1	1	CuSTRM	r	-1	-1	0			
18	PadR	255	17	0.6	20	1	1	1	CuSTRM	r	-1	-1	0			
19	Trace_Left	103	30	4	24	1	1	1	CuSTRM	r	-1	-1	0			

Figure 6.7 The "Set Loads" menu, the heart of the model.

The figure above the Set Loads menu is a copy of the trace being modeled. The numbers on this figure correspond to the lines in the Set Loads menu.

The PadL and PadR traces are the sense lines. They start at 105,17 and 255,17, respectively. They are 0.6 mm wide and extend upward 20 mm. They are 250 mm apart, almost 6 inches. The Trace_Left and Trace_Right traces simulate the magnet wires soldered to the IPC test future. Finally, we have added a thermocouple (TC Middle) at the midpoint of the trace. This computer-defined point will report the model temperature at that point in the final model results. (Later we can add multiple thermocouples to a model to track any thermal gradients that might exist.)

Finally, Figure 6.8 shows the "Run Test" menu. Most of the entries here are details, but one entry of note is that the "Temperature Dependent" box is checked. This tells the model to run several loops. After each loop the resistivity of the conductor is recalculated for the new temperature and the model runs again. Thus, this is an iterative process. The model gets closer to a stable result with each loop. Experience shows that three loops is sufficient at lower end temperatures (40°C to 50°C) and four loops is sufficient for temperature up to about 100°C.

After the model runs, a large number of files are generated. In this case the total number of files was 120! A copy of the output folder is shown in Figure 6.9. Four entries are circled. The file report.txt is a text file that summarizes all the setup and simulation data (ADAM-Research can reconstruct the simulation model from this one file alone). It contains a large amount of data, including the final temperature at the thermocouple, which is the number we want. That temperature is 65.5°C, which represents a change in temperature of 45.5°C from the ambient. The circled file "tcon01.png" is

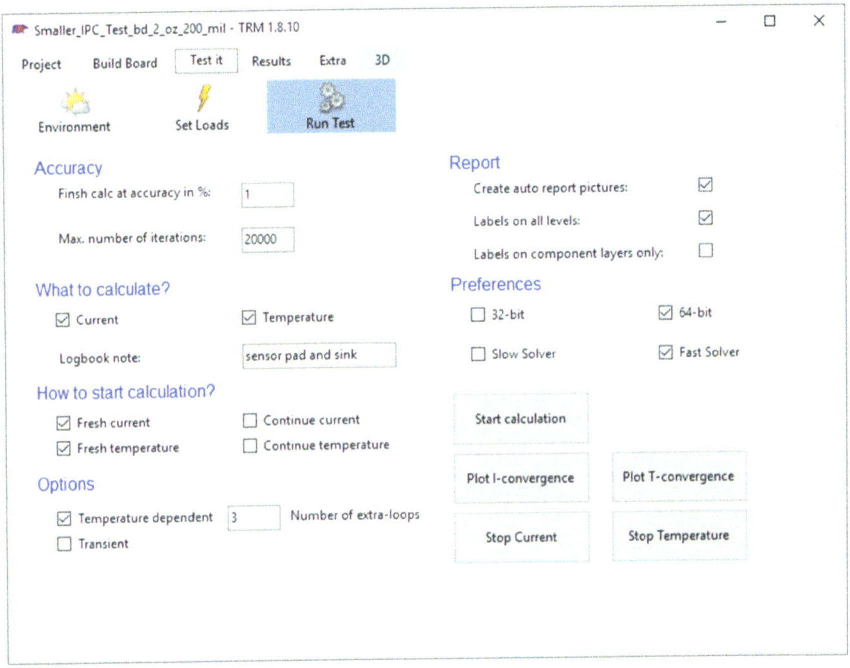

Figure 6.8 The "Run Test" menu.

the image at the top of Figure 6.7. The circled file "temp01.png" is thermal image in Figure 6.10. The folder named "fields" is filled with another large set of data files that are used for subsequent postprocessing by the TRM software.

Figure 6.10 is the thermal image file (temp01.png) for the simulation.

Several other simulations were run on the 2.0-oz IPC model. Their results are shown by red dots in Figure 6.11. The result of the specific simulation described here is circled in red. Simulations were also run on 3.0-oz external traces, which are shown in Figure 6.12. Since there is no pure IPC chart for external 1.0-oz traces, Figure 6.13 plots the simulation results against only (5.5). Running a simulation for every point on every curve is not practical, but the wide range of these results should provide confidence that the equation and the thermal simulations are valid.

One point in particular should be emphasized: *In all three sets of external curves, there is one single equation (5.5) and one single TRM model (differing only in trace width, trace thickness, and HTC).*

Simulation models were also run on selected internal and vacuum IPC traces. The results from those simulations are provided in Appendix C.

There is one item of particular note regarding traces in a vacuum. As mentioned above, the two HTC components are convection and radiation.

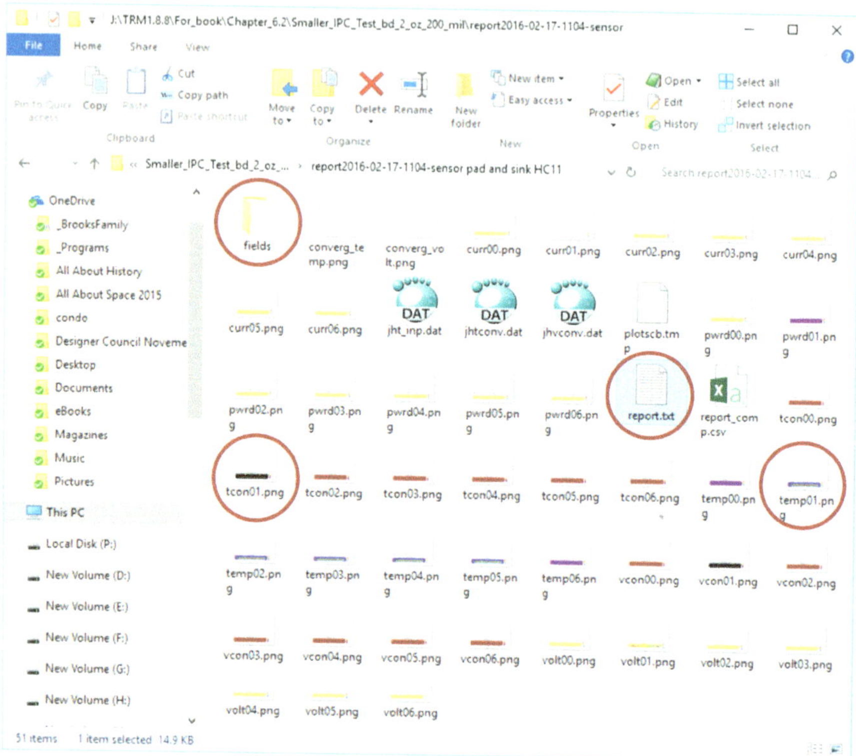

Figure 6.9 The output folder for our model simulation.

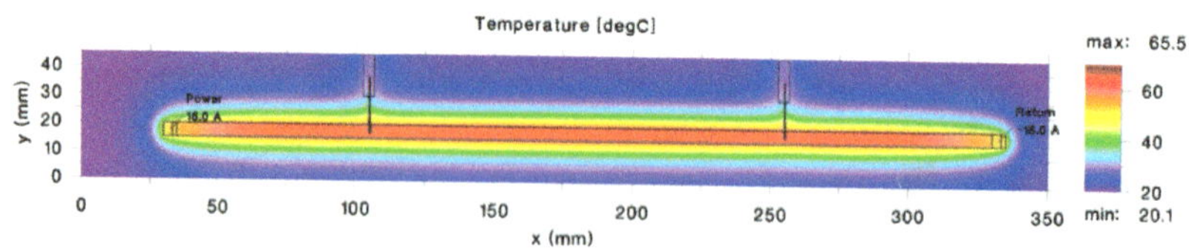

Figure 6.10 Thermal image of the IPC test board.

Very roughly speaking, they are approximately equal under laboratory conditions like this. For simulations of traces in a vacuum, the HTC values were in the range of 5 to 9, approximately half that for simulations of traces in air. This is entirely consistent with expectations.

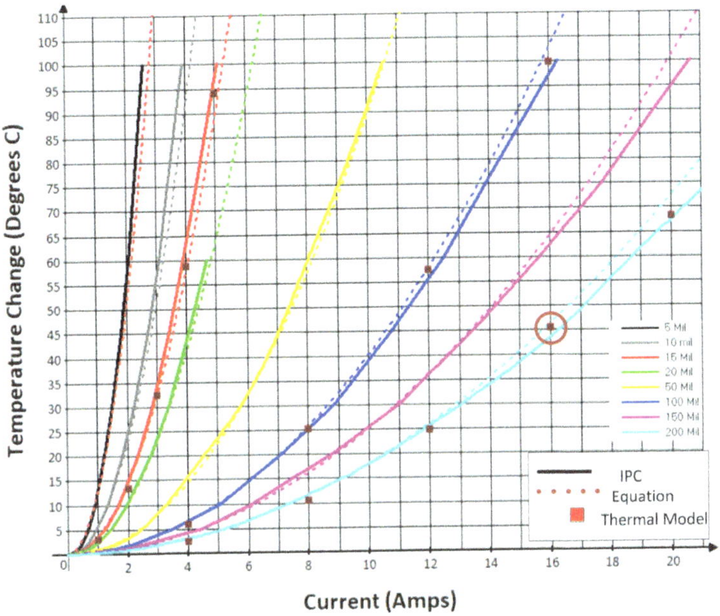

Figure 6.11 IPC 2.0-oz external curves including results of thermal simulation models.

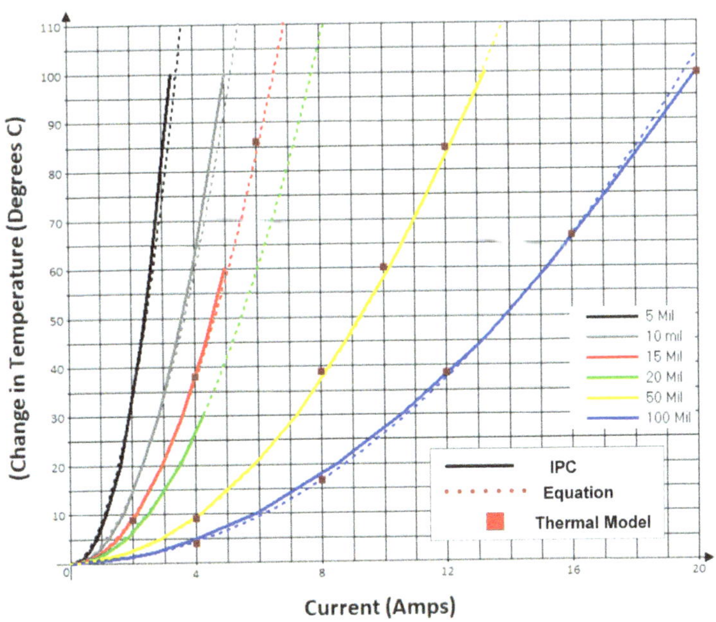

Figure 6.12 IPC 3.0-oz external curves including results of thermal simulation models.

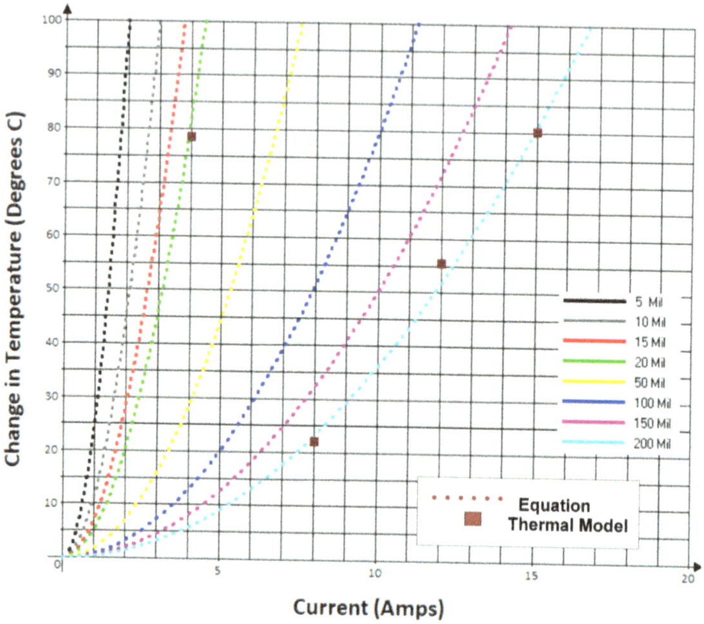

Figure 6.13 1.0-oz IPC external curves derived from (5.5) with thermal simulation results.

End Notes

1. Actually, the word "cube" is a misnomer. It is really a square column, the height of a layer. Nevertheless, we will continue to use the word "cube" through this book.
2. Units are watts-per-meter-squared-per degree Kelvin. If you would like to know more about heat transfer coefficients, start here: http://en.wikipedia.org/wiki/Heat_transfer_coefficient.

CHAPTER

7

Contents

Thermal Simulations

7.1 Bottom Line

- Simulation tools allow us to do sensitivity analyses of trace current/temperature relationships to a wide variety of variables and parameters.
- Equations and charts tend to give us worst-case values (i.e., highest temperature). Almost anything we do from a real-world board design standpoint tends to lower the trace temperature, although it might tend to raise the temperature of the dielectric around the trace.
- Trace temperatures are very sensitive to trace and board thickness, copper resistivity, and dielectric thermal conductivity coefficients.
- Internal traces are cooler than external traces and this effect begins at very shallow depths.

7.2 Sensitivities: Layout Parameters

A simulation tool such as TRM allows us to simulate much more varied and complex trace structures than we can, from a practical standpoint, measure in a laboratory environment. In this section we will look at a variety of board layout parameters under the board designer's control and see how the thermal results might change as we change those parameters. In the next section we will look at the sensitivities of trace temperature to material property variations that the designer may not be aware of or have control over. In the final section we will look at the sensitivities of trace temperature to the position (depth) of internal traces.

7.2.1 Small Trace Widths

The current/temperature curves are very steep for narrow traces. Figure 7.1 illustrates the curves for 1.0-oz, 5- and 10-mil wide internal and external traces. They are based on the equations from Table 5.1.

For 5-mil traces the difference between a 500-ma current and a 1.0-amp current is about 20°C. A 1.0-amp current has about a 25°C temperature increase, while a 2.0-amp current has an almost 100°C increase! The situation is only slightly better for 10-mil wide traces. As soon as we start carrying any significant currents at all, the trace temperatures increase quickly.

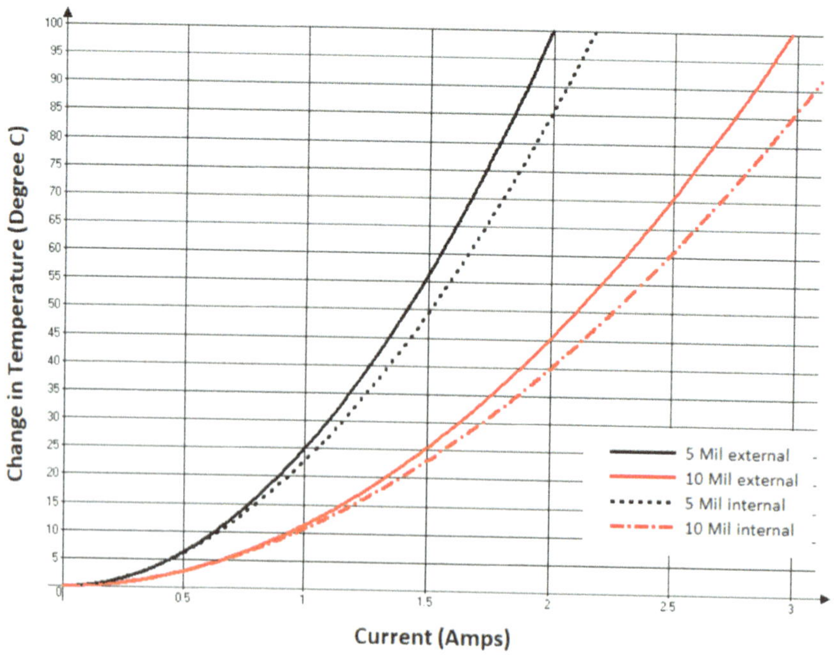

Figure 7.1 1.0-oz, 5-mil, and 10-mil external and internal curves.

Table 7.1
Trace Temperature as a Function of Trace Length

Test Condition	Center Temperature (°C)
6-inch trace measured	67.0
6-inch trace, simulated	69.2
4-inch trace, simulated	68.0
2-inch trace, simulated	60.9
2-inch trace, simulated, plane underneath pads	59.3
1-inch trace, simulated	49.0
1-inch trace, simulated, plane underneath pads	46.3

Figure 7.2 shows the 5- and 10-mil wide external traces, compared to 4- and 9-mil traces, respectively. Depending on how great the fabrication tolerances are, unexpectedly high trace temperature variations might develop.

The bottom line is that if trace temperature for very narrow traces is a concern, designers might want to be particularly conservative in specifying trace widths.

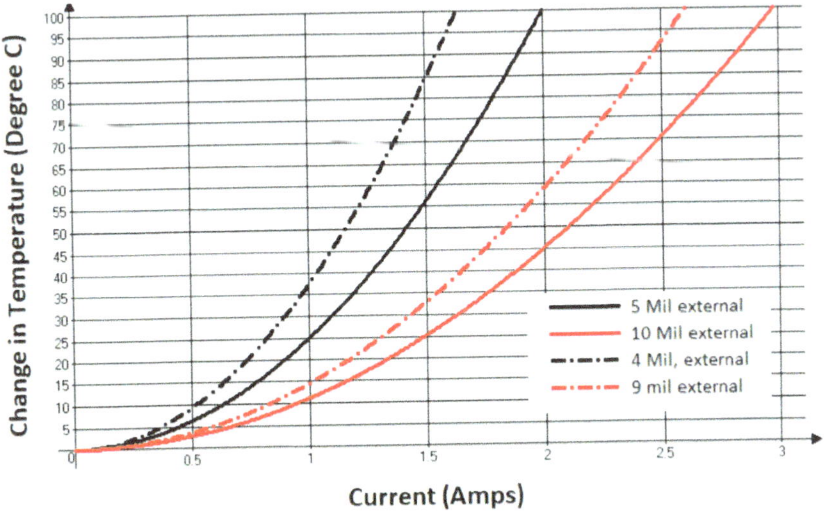

Figure 7.2 Comparing possible fabrication tolerances for 5- and 10-mil external traces.

7.2.2 Transient Response

Traces do not heat immediately. They take some time to reach temperature, sometimes a surprisingly long time. This can be a problem for people running laboratory tests of trace current/temperature relationships; they must be sure they are waiting long enough for the trace temperatures to stabilize.

Using a simulation model for a 1.0-oz, 6-inch long, 200-mil wide external trace driven at 15 amps, Figure 7.3 shows a curve of how long it takes the trace to reach a stabilized temperature of 94.6°C. It takes almost 3.5 minutes for the trace to get within 90% of its final value, almost 5 minutes to get within 95% of its final value. A curve for an internal trace rising to the same temperature is also shown as the red dotted curve. The internal trace rises slightly more slowly than does the external trace because internal traces cool more efficiently than do external traces. Therefore, it takes longer for internal traces to reach their stabilized temperatures.

7.2.3 Thermal Gradients

For this and some of the analyses to follow, we had access to a test board to check measurements experimentally against a simulation model (more about where the test board came from in Chapters 8 and Appendix B). So we put together a standard simulation model that matched some of the

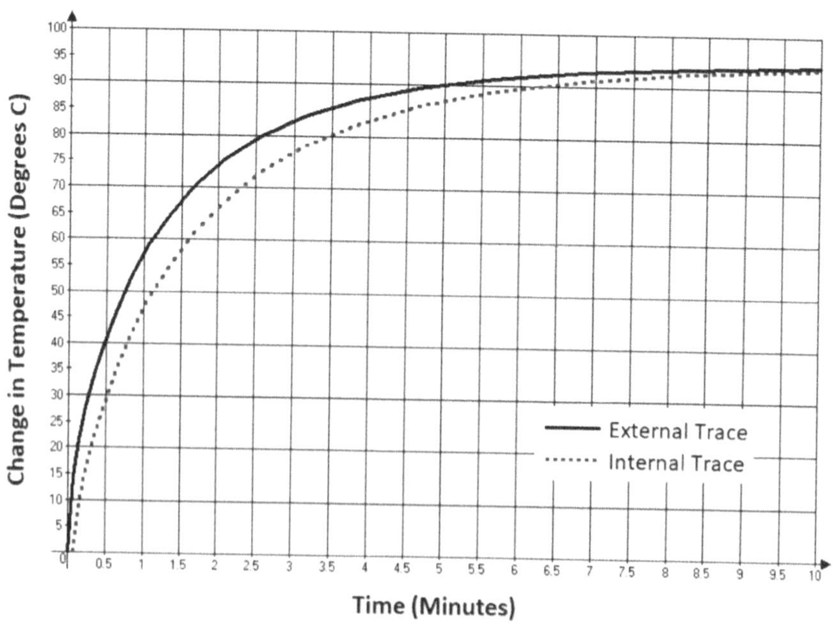

Figure 7.3 Heating time for a 1.0-oz, 200-mil wide trace carrying 15 amps.

parameters of the test board for analysis. Figure 7.4 illustrates a diagram of that model.

The board outline is 192 mm × 25 mm. There are copper strips along the top and bottom edges to simulate the adjacent traces on the test board. The pads (left and right) are 0.5 mil × 0.3 mil (13 × 7.8 mm). The trace length is 6 inches (152 mm). The trace itself is centered on the centerline of the board. Its width is adjusted as necessary for the particular simulation. A second copper strip is added (not shown) to simulate a parallel trace. Trace thickness is nominally 1.9 mils (approximately 1.5 oz). The board thickness is 63 mils (1,600 μm). Board thermal conductivity was measured by C-Therm Technologies (see the Acknowledgments and Appendix A) as 0.68 W/mK (in-plane) and 0.51 W/mK (through-plane). This represents the actual traces we had available for experimental testing. This simulation model was then modified as necessary for additional simulations (different trace lengths, additional planes, etc.).

There is a clear thermal gradient from the center of the trace (usually, but not always, the highest temperature point) out to the end of the trace. A typical simulated thermal gradient is illustrated in Figure 7.5. This is for a simulated 100-mil wide, 6.0-inch long trace carrying 9.35 amps. The maximum temperature of the trace is 69.2°C.

Our test board contained traces with widths of 27, 100, and 200 mils. We were able to measure the actual thermal gradients of each of those traces and compare them with simulated version of the same traces. The results are shown in Figure 7.6. Notice how sharply the thermal gradients drop off as you get closer to the ends of the traces.

7.2.4 Changing Trace Length

Our standard test trace length is 6.0 inches. It is interesting to see how sensitive the temperature is to trace length. Using the 100-mil wide simulation, we ran simulated test lengths of 6.0 inches, 4.0 inches, 2.0 inches, and 1.0 inch, as well as 2.0 inches and 1.0 inch with a heat sink under the pads (a simulated plane on the opposite side of the board. The plane did *not* extend

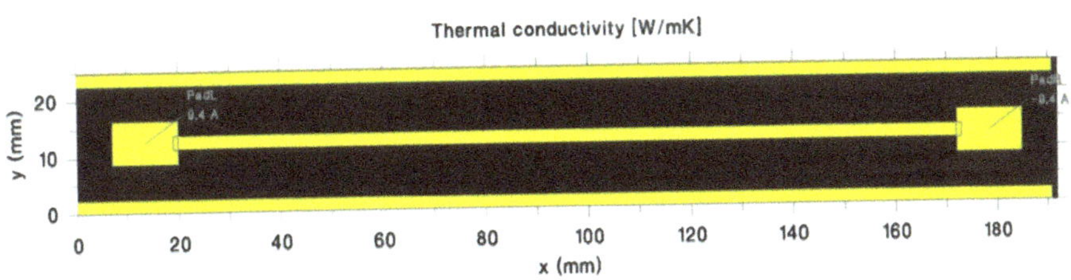

Figure 7.4 Standard simulation model.

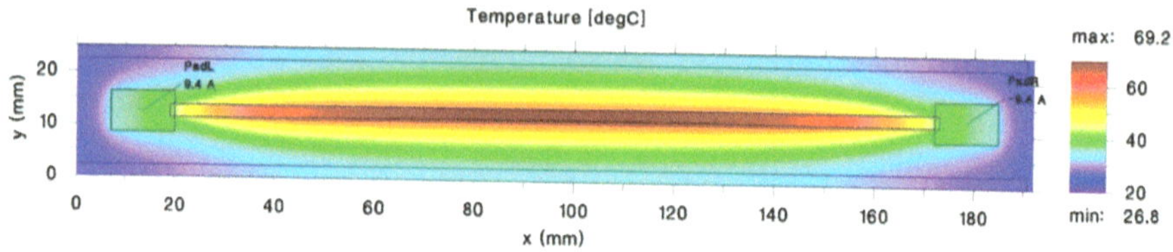

Figure 7.5 Thermal gradient for a 6.0-inch, 100 mil wide trace carrying 9.35 amps.

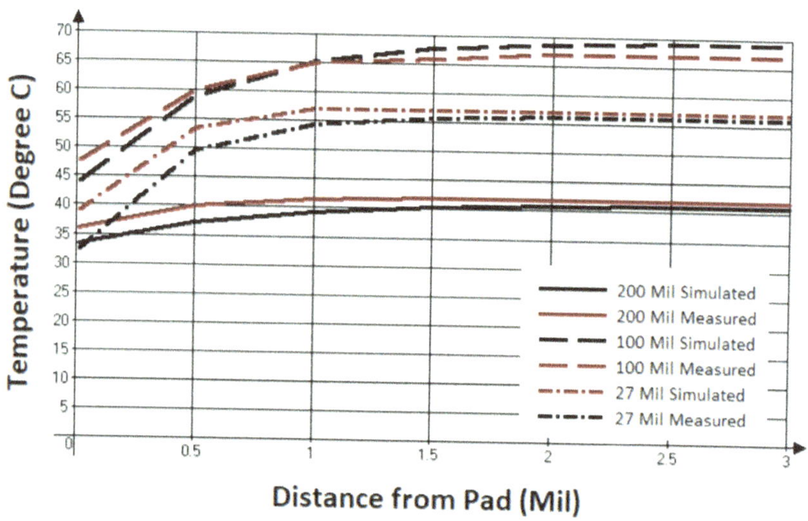

Figure 7.6 Measured and actual gradients for three trace sizes.

under the trace). Current was 9.35 amps in all cases. We could also measure an actual trace that was 6 inches long. The results are summarized in Table 7.1.

There is a clear tendency for the temperature to drop as the trace shortens, but it is not a particularly strong tendency unless we are dealing with very short traces. The size of the pads has an influence on the degree of temperature drop.

7.2.5 Dimensional Uncertainties

While designing the test board, we specified the nominal trace thickness for the traces at 1.9 mils, (1.0-oz film coated with 0.5-oz deposited copper). The 100- and 200-mil traces were located at one end of the board, while the 27-mil trace was located more toward the other end. We noticed a discrepancy in some of the data regarding the 27-mil trace (and those near it), so we went back and microsectioned the board to verify trace thicknesses. We found that the thickness of the 27-mil trace was actually more like 2.7 mils (almost 2.0 oz)! This was a difference of around 40%. In fact, that is why the measured temperature for the 6.0-inch trace in Table 7.1 is lower than the simulated temperature.

Such differences are not uncommon on production-type boards. It is axiomatic that you cannot take the nominal trace dimensions as given when doing analyses. In fact, this is one of the larger problems in doing experimental analyses of boards—knowing the *actual* dimensions you are dealing with. We stressed this point back in Section 3.5.

Table 7.2 is a comparison of the simulated results based on the two trace thicknesses for a 27-mil wide trace carrying 4.75 amps.

7.2.6 Presence of Planes

Most of today's boards have planes in them. The planes play a significant role in cooling the traces above them, because the planes offer a significant heat-conducting path away from the trace. Earlier we looked at our standard model of a 100-mil wide trace with no plane. In Figure 7.7 we see what happens when we add a plane to the board; first to the bottom layer of the board, and then 10 mils underneath the trace layer. The temperature ranges from 69.2°C to 54.1°C to 45.9°C, respectively, for the three cases. The results are tabulated in Table 7.3.

But even more importantly, look at the thermal halo around the traces. For the board with no plane, most of the heat is concentrated directly under and around the trace. As a plane gets closer to the trace layer, the surrounding area heats up and the halo spreads out dramatically wider. Thus, planes help cool the particular trace of interest, but they may spread the heat more efficiently to adjacent traces, with uncertain consequences.

7.2.7 Adjacent Trace

In most practical boards, real estate is valuable, and so there is almost always a trace nearby the trace we are concerned about. In this model we

Table 7.2
Impact of Dimensional Uncertainty

	Thickness	
	1.9 Mil (Nominal)	2.7 Mil (Actual)
Temperature with 4.75 amps (°C)	77.8	56.2

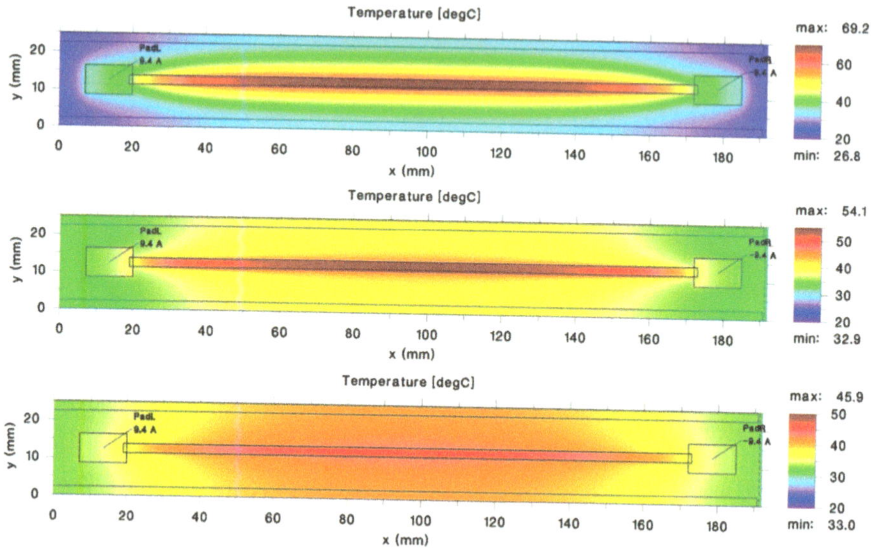

Figure 7.7 Impact of planes: no plane (top), plane on bottom layer (middle), plane 10 mils under traces (bottom).

Table 7.3
Impact of Underlying Plane on Trace Temperature

Plane Position	Trace Temperature (°C)
No plane	69.2
Bottom of board	54.1
Under trace	45.9

will add another 100-mil wide trace separated from the trace of interest by 8 mils. A diagram of this configuration is shown in Figure 7.8.

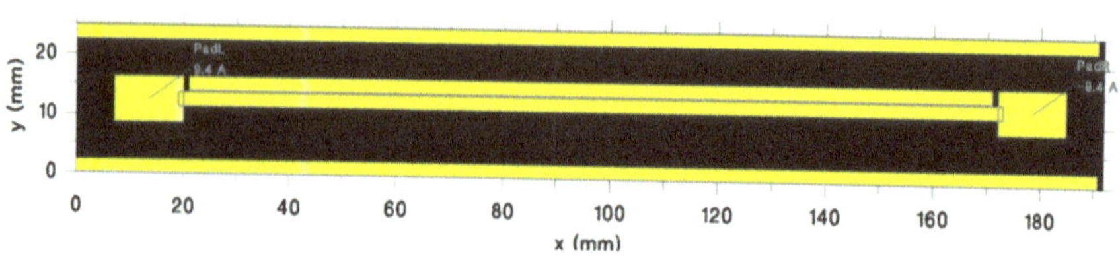

Figure 7.8 Adding an adjacent trace.

The result of this model, compared to the trace without a second adjacent trace, is shown in Figure 7.9. The temperature of the current carrying trace (9.35 amps) drops from 69.2°C without a parallel trace, to around 65°C when there is a parallel trace. However, the temperature of the parallel trace rises to around 55°C! Thus, the temperature rise of one trace may have significant implications for traces surrounding it on the board.

7.2.8 Adjacent Trace with Underlying Plane

When we added a plane underneath the trace (Section 7.2.6), the temperature of the trace dropped to 45.9°C. When there was an adjacent trace (previous section) the adjacent trace heated to about 55°C. Now if we add an inner plane 8 mils under this structure, the temperature of the driven trace becomes 45.7°C and the temperature of the parallel trace drops to 43.0°C (See Table 7.4). Thus, the biggest impact of adding a plane under the parallel structure is to help safeguard the second, parallel trace.

7.2.9 Parallel Power Traces

Designers sometimes ask about what happens if we split the power trace into two traces. Figure 7.10 illustrates this. Instead of a single, 100-mil wide trace carrying the current, we have two 50-mil wide traces (separated by 50 mils) carrying the current. Since the current splits evenly between two

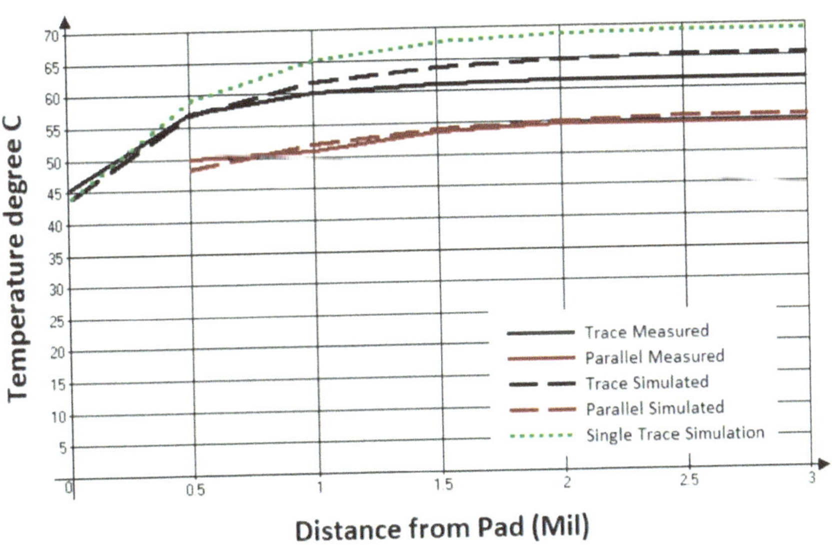

Figure 7.9 Effect of a parallel trace.

Table 7.4
Adding a Parallel Trace and an Underlying Plane

Condition	Driven Trace	Adjacent Trace
Single trace	69.2	—
Single trace over plane	45.9	—
Parallel trace	65.0	55.0
Parallel trace over plane	45.7	43.0

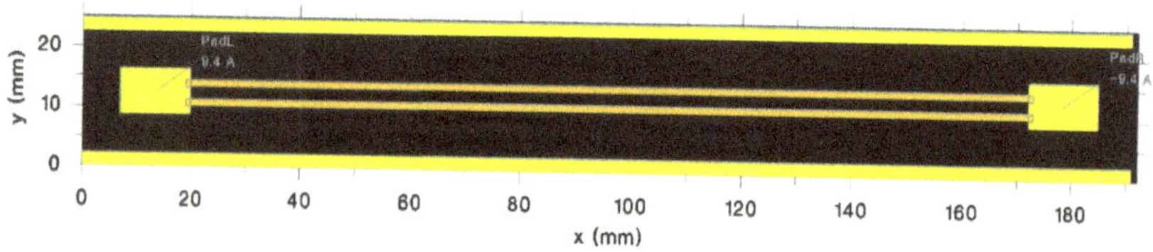

Figure 7.10 parallel traces carrying the current.

traces (each half as wide as before), the current density in each trace is the same as before.

The maximum temperature in each trace is 64.7°C, slightly lower than before. But the temperature of the board area between the traces rises to 59.1°C. Figure 7.11 illustrates the thermal image that results. Compare this image with Figure 7.5. As a general rule, splitting a single power trace into two equal traces, each half as wide, does not buy you anything. The temperatures are nearly the same and the thermal envelope is wider.

7.2.10 Stacked Power Traces

Instead of splitting the power trace into two parallel traces (Figure 7.11), we split them into two traces stacked one over the other. The results of that model are not shown here, but they are almost indistinguishable from the single trace model discussed above and shown in Figure 7.5. The

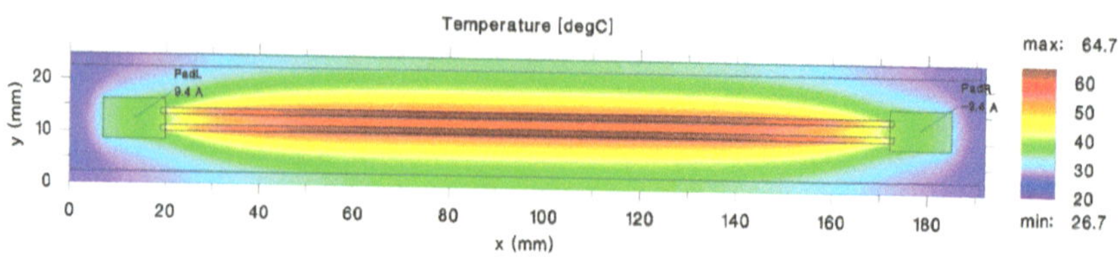

Figure 7.11 Thermal image of two parallel traces carrying the current.

temperature is the same, 69.2°C, and the thermal profiles are almost identical. This would be expected because even though we have two traces instead of one, each generating half the heat, we are generating the same total amount of heat in almost the same volume of material.

So our conclusions remain the same: As a general rule, splitting a single power trace into two equal traces, each half as wide, does not buy you anything. The temperatures and thermal envelopes are nearly the same.

7.2.11 Air Flow

Most of the simulations done in this chapter were done with HTC set to around 12 to 14. Experimental calibration seems to suggest that this is a reasonable value. If we blow air across the trace, the effect is to increase HTC. But how to equate air flow to HTC when dealing with an isolated, small trace is extremely difficult to estimate.[1] Just for fun, we took the basic simulation model for the 27-mil trace and increased the HTC to 20 and to 28. These values would simulate supplemental cooling, but how much and of what kind would be very uncertain. Therefore, the results are obviously to be taken with a large grain of salt. The results are shown in Table 7.5. It should be noted that HTC variations have greater impacts at higher temperatures.

7.2.12 Summary

Wow! Talk about confusing. That was a lot of information. Table 7.6 is an attempt to summarize it all.

But when we are done, what does it all mean? The bottom line is that this topic is very complicated and there simply are no easy answers. But there are a few generalizations we can make:

- It is almost universally true that variations in results increase with temperature. That is, at low temperatures (associated with lower currents) the various parameters we have looked at here make little or

Table 7.5
Impact of Changing Heat Transfer Coefficient

HTC	Temperature
14	56.2
20	51.1
28	46.9

Table 7.6
Summary of Sensitivity Simulations

	Temperature (°C)	
Simulation Model	Basic Trace	Parallel or Adjacent
6 inch	69.2	
4 inch	68.0	
2 inch	60.9	
2 inch with sink*	59.3	
1 inch	49.0	
1 inch with sink*	46.3	
6 inch-gradient†	44.8	
With opposite plane	54.1	43.3§
With underlying plane	45.9	44.7§
Parallel trace	62.0	55.0¶
Parallel + plane	45.7	43.0¶
Split power trace	64.7	59.1#
With air flow‡	46.9	

*Heat sink under pads.
†Measured adjacent to pad.
‡Very approximate.
§Adjacent plane maximum temperature.
¶Adjacent trace maximum temperature.
#Dielectric between traces.

no difference. As the temperature (current level) increases, changing these parameters has a correspondingly greater effect on the outcome.

- The IPC curves generally represent a worst case scenario. Any other variation we introduce lowers the trace temperature, sometimes considerably so.
- We have not addressed some of the more complex shapes, such as the fillets in thermal reliefs or the copper plating in drill holes. While we can speculate that the I^2R heating in these cases might be similar to other simulations, the cooling situations might be substantially more complicated.
- The relationship between trace currents and temperatures is *very* complex. It is too complex to model with a single set of equations or curves. Since board real estate is very expensive, board designers usually want to use the smallest traces their fabricators will allow while still meeting the requirements. Optimizing board area when considering thermal effects does not seem possible without sophisticated thermal simulation software.

7.3 Sensitivities: Material Parameters

In the previous section we looked at the sensitivity of the current/temperature relationship to various layout parameters. In this section we look at the current/temperature sensitivities to various material parameters and properties. To do so, we will develop a standard computer model to start with. That model is a 100-mil wide, 1.0-oz thick, inch long trace on an FR4 dielectric (board) measuring 6.5 inches long by 2 inches wide. This dielectric is 63 mils thick, which is typical for modern PCBs. It will carry 7 amps. The following parameters will be u sed in the base model. They have been found by the authors in previous studies to be typical and reliable.

Resistivity: 1.68 μohm-cm

HTC: 11

FR4 with thermal conductivity coefficient

In-plane 0.6 W/mK

Through plane 0.4 W/mK

Seven amps of current is enough to raise the trace temperature to almost 60°C, 40° above ambient. As noted above, sensitivities to layout and material parameters is a function of temperature increase, so this is a reasonable place to start. We will then vary one parameter/property by itself and look at its effect on trace temperature. In effect, we are looking at the partial derivative of the current/temperature relationship with respect to that parameter.

7.3.1 Board Thickness and Planes

As the thickness of the dielectric increases, there is more board material for the heat to conduct through, so the temperature will presumably be lower with increased thickness. This is true up to a point, but there reaches a point where increased thickness does not buy you much. The typical PCB board is around 63 mils thick, and this is the thickness used for the basic model. But boards can be much thicker than that. Figure 7.12 illustrates the trace temperature as a function of board thickness. Note how the curve starts out fairly steep for thin boards, but then tends to stabilize as the board thickens.

The dynamics associated with this effect can be seen in the thermographs in Figure 7.13 and 7.14. These show the temperature of the board material at and below the trace. Imagine a surface cutting through the middle of the trace (at the 80-mm point in the board) along the *x-z* plane. The view is toward the cross section of the trace and shows the temperature

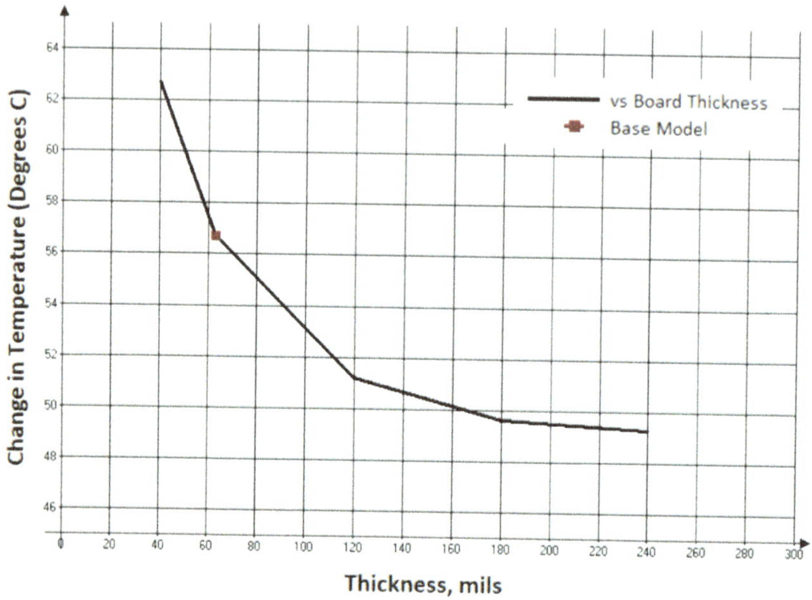

Figure 7.12 Trace temperature decreases as the board gets thicker, up to a point.

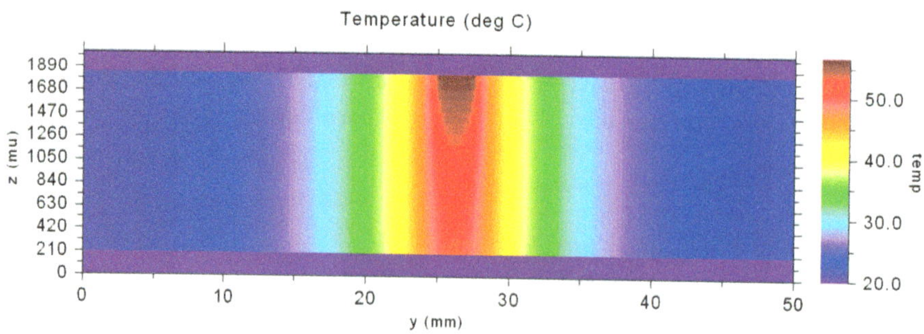

Figure 7.13 Vertical thermal profile of the 63-mil thick board.

gradient from the trace down toward the bottom of the board material. Figure 7.13 is for the 63-mil thick board, and Figure 7.14 is for the 240-mil board. There is not enough material in the thinner board for the heat to spread out much, but that is quite different for the thicker board, which is why the trace on the thicker board is significantly cooler.

Perhaps the largest external impact on trace temperatures can come from an adjacent plane. Many modern boards have internal planes, especially those with signal integrity issues. We did one simulation with a 63-mil thick board with two internal planes, one 20 mils under the trace, and the

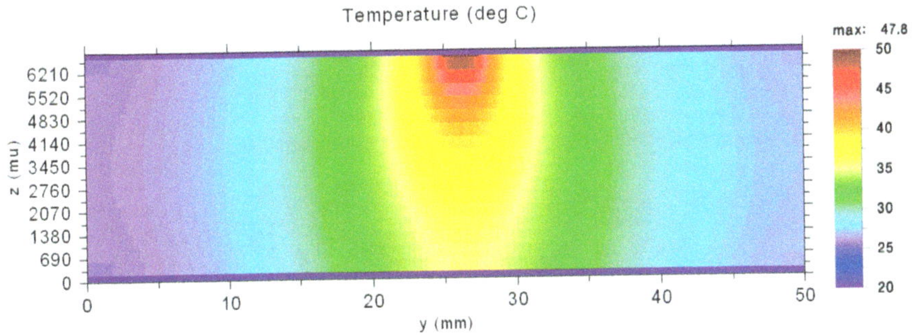

Figure 7.14 Vertical thermal profile of the 240-mil board.

other 40 mils under the trace. These planes covered the entire surface area of that layer.

The impact of a plane is significant. It dropped the base temperature from 56.7°C to 34.5°C. But more than that, it spread the heat (and temperature) out much more widely than was the case without the plane. Figure 7.15 compares the thermograph of the trace layer with the underlying plane (b) to that of the trace without a plane (a). There is a dramatic difference.

The vertical distribution of the temperatures is also dramatically different. Figure 7.16 shows the vertical thermal profile for the case with the planes. Compare it with the case without a plane (Figure 7.13).

7.3.2 Effect of Resistivity

The resistivity of pure copper (at 20°C) is 1.68 μohm-cm. If the copper material on the board is plated (ED) copper, it will have a resistivity very close to this value.[2] Rolled copper, on the other hand, probably originates as a copper alloy, and has a resistivity anywhere between 1.72 and 1.78 μohm-cm. Copper foil on PCBs can be either ED or rolled, but rolled is more common. The relationship between resistivity (p) of the material and the resistance (R) of the conductor is based on the geometry (area A and length L) of the conductor:

$$R = (\rho / A) * L \tag{7.1}$$

Recall that the heating of the conductor is based on I^2R power dissipation in the conductor. Figure 7.17 shows the relationship between the trace temperature and the resistivity of the copper. Our standard model assumed pure copper, and therefore the temperature goes up as resistivity varies from that. Note that the curve is linear, as would be expected.

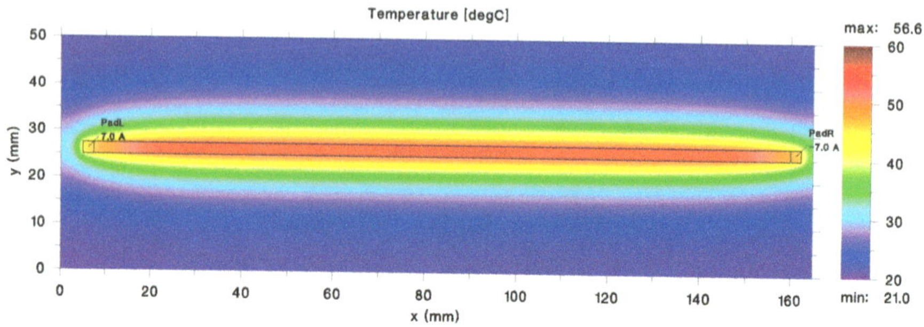

(a) No plane

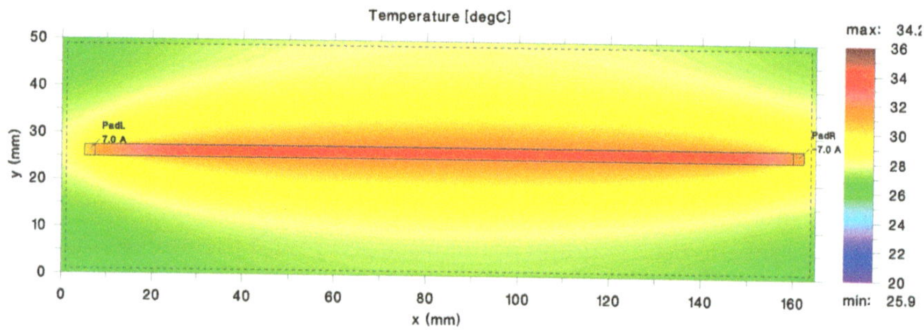

(b) Plane under trace

Figure 7.15 Impact of a plane on the thermal heat spreading.

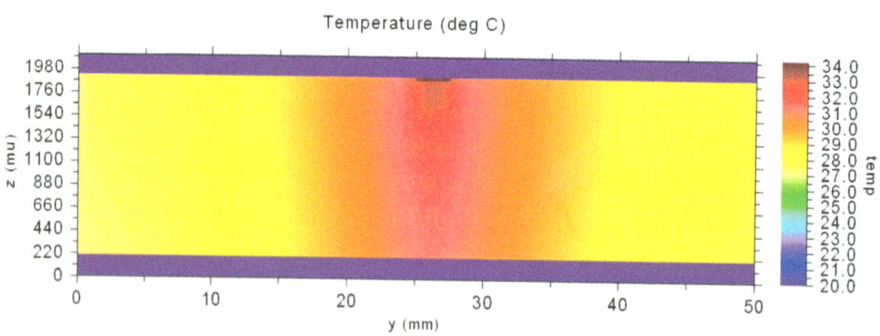

Figure 7.16 Vertical thermal profile with planes.

7.3.3 Effect of Heat Transfer Coefficient

As mentioned above, HTC has components of convection and radiation. The authors have found in previous studies that for this temperature range, an HTC of 11 is pretty representative. An HTC of 6 would represent the situation of a trace in a vacuum (e.g., space). HTCs above 11 would represent

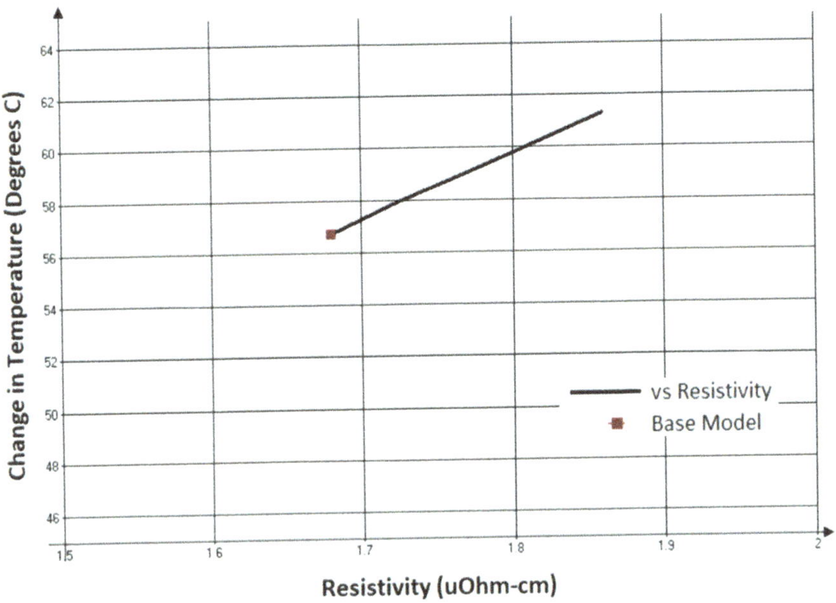

Figure 7.17 Trace temperature as a function of resistivity of the conductor.

some form of supplemental cooling, perhaps a heat sink or blowing air. We have looked at a range of HTC from 6 (vacuum) to 18, but there is no practical way today to say what 18 actually means! Figure 7.18 shows the relationship between trace temperature and HTC. The relationship shown in the figure is not surprising; as the convection cooling increases, the trace temperature goes down, sometimes dramatically.

7.3.4 Effects of Thermal Conductivity Coefficient

Start with the idea that the board material has the greatest impact on trace cooling, and then with the idea that the thermal conductivity coefficient is the measure of how effective the board is at cooling the trace. Then recognize that this measure—the thermal conductivity coefficient—is rarely provided in the data sheets of the material suppliers! And if it is provided, it is frequently provided as a single value, without specifying which coefficient it is, or it is often the value (through-plane) that is least important. This is an unfortunate state, indeed.

Figure 7.19 shows the relationship between the thermal conductivity coefficient and the trace temperature. The in-plane (*x-y*, or horizontal) coefficient is clearly the one that has more effect on trace temperature[3]

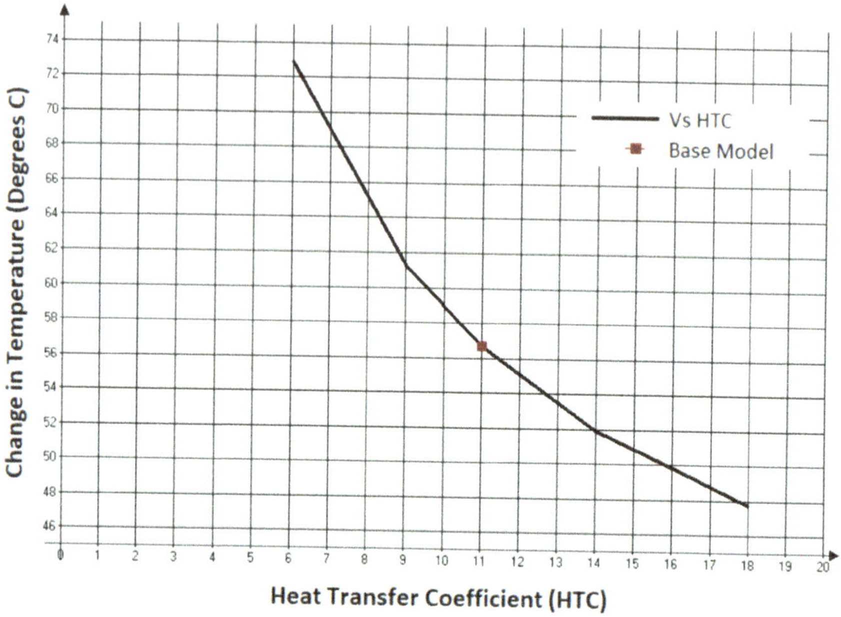

Figure 7.18 Trace temperature as a function of HTC.

7.3.5 Effect of Trace Thickness

Trace thickness is, of course, directly related to cross-sectional area. Therefore, a change in thickness will have a significant effect on the current/temperature relationship. But some people might be surprised by how big an effect it can have.

Figure 7.20 shows the relationship. The red dot in the figure shows the results of the basic model, using a 1.0-oz thick trace. The curve of temperature versus thickness is moderately steep in this range. For this model, the sensitivity is about 1.0 degree C for a change of 0.03-mil thickness. This is likely one reason traces do not heat uniformly (as discussed in Chapter 13). Normal variations in trace thickness are well within this range.

7.3.6 Summary

The results of the simulations in Sections 7.3.2 through 7.3.5 are summarized in Table 7.7. The values in the table, in effect, represent the partial derivatives of the change in temperature as a result in the change of one of the variables. While all the values are significant, perhaps the most sensitive parameter is thickness. We discuss the influence of thickness in more detail in Chapter 13.

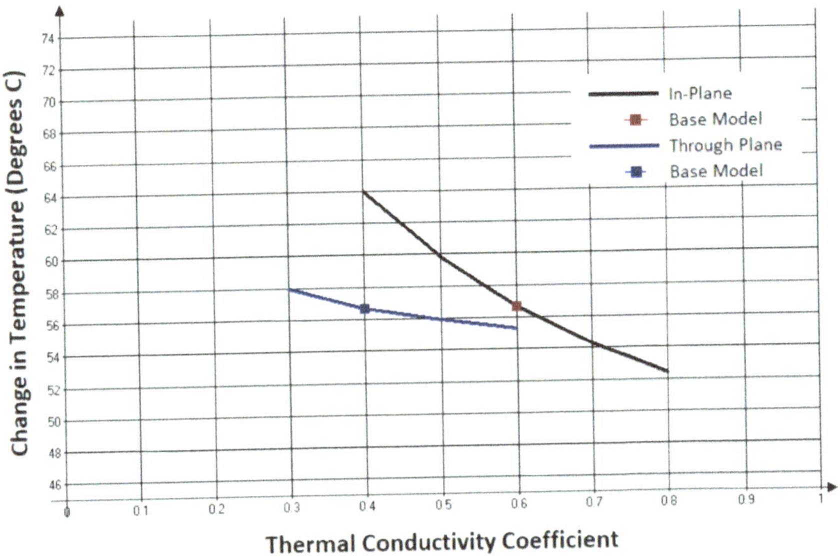

Figure 7.19 Trace temperature as a function of thermal conductivity coefficient.

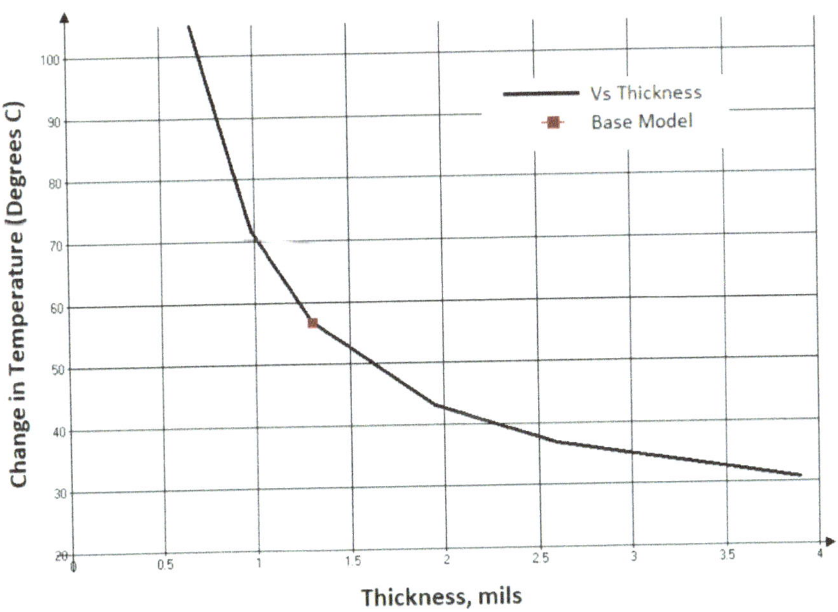

Figure 7.20 Trace temperature as a function of trace thickness.

Table 7.7
Impact of MATERIAL Parameters on Trace Temperature

Change in Temperature (°C)	For Each
1.0	2% Δ resistivity
3.0	0.1-mil Δ thickness
5.0	HTC between 11 and 14
5.6	Thermal conductivity coefficient, 0.5 to 0.7

7.4 Sensitivities: Trace Depth

An interesting question is, "Is there a predictable relationship between the external change of temperature of a trace and the internal change of temperature of the same trace carrying the same current?" That is, is the internal trace change 10% cooler, or 20% cooler, or is there some other predictor we can use to predict internal changes in temperatures? The data in IPC-2152 make it possible to explore this question. Internal and external temperature change curves are provided therein for 1.0-, 2.0-, and 3.0-oz traces and a variety of trace widths, so these relationships can be explored directly.

Using the data in IPC-2152, we can specify a trace width, trace thickness, and current level, and then read off from the graphs the corresponding internal and external changes in trace temperatures. We can do this for a sufficiently large number of points to generate a reliable regression analysis. Let us define three terms for this analysis:

1. *DTi* is the change of temperature for an internal trace at any depth;
2. *Dte* is the change of temperature of an external trace;
3. *Dti*(min) is the change of internal trace temperature at the middle of the board, where the internal trace temperature is the lowest.

The regression analysis results in a relationship between the two changes in temperature (Dti vs Dte) that was found to be much more complicated than first assumed. The regression analysis resulted in (7.2):

$$DTi = 0.82 * Th^{.053} * W^{.021} * DTe^{.976} \tag{7.2}$$

where

DTi = change in temperature, internal trace, °C

DTe = change in temperature, external trace, °C

TH = trace thickness in mils

W = trace width in mils

The results illustrate that there is indeed a predictable relationship between the external temperature of a trace and the internal temperature of the same trace carrying the same current. And that relationship is a function of the external trace change in temperature. The greater the external trace change in temperature is, the greater the difference between the internal and external trace temperatures will be.

Since (7.2) is nonlinear and involves wide ranges in the values of the variables, it is difficult to represent it graphically and to draw any conclusions from it. But we can normalize it by dividing DTi by Dte. That gives us the ratio, Dti/Dte, which is more intuitively useful. This ratio is the ratio of the internal change of temperature to the external change in temperature as a function of the external change of temperature. Doing so results in (7.3):

$$DTi/DTe = 0.82 * TH^{.053} * W^{.021} * DTe^{-.024} \tag{7.3}$$

Examining (7.3), we can determine a couple of relationships:

1. The ratio of internal change of temperature to external change of temperature decreases with increasing external trace temperatures, all other things being equal. That is, internal and external temperature differences are relatively small at lower temperature changes. But internal traces get relatively cooler as (external) temperatures increase. This is because internal traces cool more efficiently than external traces do.
2. Thinner (internal) traces cool more efficiently than do thicker traces, all other things being equal.
3. Wider (internal) traces cool more efficiently than do narrower traces, all other things being equal. The reason for both these effects is that wider and/or thinner traces have relatively more surface area in contact with the dielectric and therefore conduct heat away from the trace more efficiently.

These three relationships can be seen in Figure 7.21.

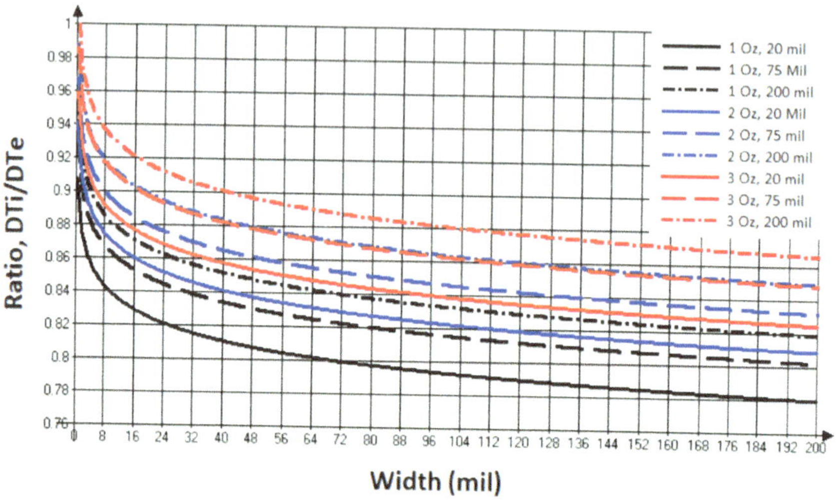

Figure 7.21 Ratio of internal change of temperature to external change of temperature for a variety of trace configurations (derived from (7.3)).

The results herein apply for internal traces in the middle of the board. Another question is, "What happens to the relationship for traces closer to an outside layer (i.e., not in the middle of the board)?" For any given trace configuration and current, the *difference* in temperature between the internal and external trace will be a minimum for an internal trace very near the surface of the board, and a maximum for a trace at the center of the board. (The *maximum difference* is what results from (7.2).)

Modeling we have done with TRM suggests that the cooling effect internal to the board kicks in at very shallow depths. That is, the difference between DTe and Dti (i.e., Dte-Dti) is very small at the surface of the board, but starts increasing sharply at very shallow depths. We can define a term *fraction* to be the ratio of the difference between Dte and Dti at any depth in the board to the maximum difference between Dte and Dti at the middle of the board. In (7.4), Dte and Dti are defined as before. Dti(min) is the minimum value of the internal change of temperature at the midpoint of the board. That is where the maximum difference in the external and internal temperatures will occur.

$$Fraction = (DTe - DTi)/(DTe - Dti(min)) \quad (7.4)$$

where *Fraction* ranges from 1.0 at the midpoint of the board (where the difference is the greatest) to 0.0 on the surface layer (at the limit, where there is no difference). We estimate the relationship to be that shown as (7.5) and plotted in Figure 7.22.

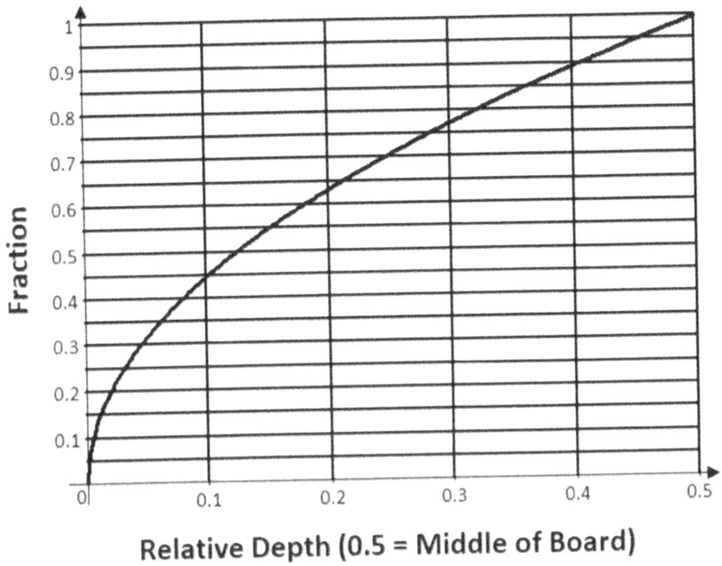

Figure 7.22 Fraction of full internal temperature difference as a function of trace depth.

$$Fraction = 1.41\sqrt{Depth} \tag{7.5}$$

where *Depth* = the depth of the trace expressed as a fraction of the total board depth (0.0 < *Depth* < 0.5).

Figure 7.22 illustrates this. At a relative depth of 0.5 (middle of the board) the ratio of the internal temperature change to the external temperature change is 100% (1.0) of the result predicted from (7.2). But let's say a trace is at a relative depth of 0.1 (that would be just 6 mils into a 60-mil thick board). At that point the difference in trace temperatures would be 42.5% of the maximum difference, DTc – DTi(min).

As an example, let's assume a 60-mil thick FR4 board with 20-mil wide, 1.0-oz traces on the top layer and at the middle of the board. If the external trace is carrying 2.8 amps, we might find that the external change of temperature is 40°C. We can calculate the internal change in temperature, DTi, from (7.2) or from Figure 7.21. Using Figure 7.21, we can calculate the internal change of temperature at the midpoint of the board, DTi(min), as $40 \times 0.81 = 32.4$°C. That is 7.6°C cooler than the external change of temperature. Now if the trace is 12 mils below the surface (instead of the middle of the board); that is, the trace is at a depth of 12/60 = 0.2, then, from Figure 7.22, the difference between the internal temperature and external temperature changes would only be 0.63 times that just calculated, or $0.63 \times 7.6 = 4.8$°C. The internal change of temperature, DTi, would be 35.2 (i.e.,

40 – 4.8) °C (a few degrees hotter than at the midpoint), but still cooler than the external trace.

We can achieve this result directly by recognizing that we are deriving the internal change of temperature from (7.2) as modified by (7.5) as follows:

$$DTi = DTe - \left(DTe - DTi(\min)\right) * fraction \tag{7.6}$$

Plugging the example above into this equation:

$$DTi(\min) = 32.4 \tag{7.7}$$

$$Fraction = 1.41*0.2.5 = 0.63 \tag{7.8}$$

$$DTi = 40 - (40 - 32.4) * 0.63 = 35.2\ °C \tag{7.9}$$

There is a caveat to this analysis. This analysis applies to the data in IPC-2152. Other situations (i.e., board materials) may result in different relationships. So this analysis is a guide, and further research is needed before we can say that it is reliable for all situations. Nevertheless, it will give the user a general idea of how trace temperatures change as we progress further into the board.

7.5 Conclusions

Many board designers need to work hard to optimize their designs. That often means designing the narrowest traces they can or that their fabricators will allow. The reason is that board area is expensive. Narrower traces mean more traces per layer, which mean fewer layers. And layers are expensive.

So if a trace is carrying a significant current, the designer wants to design the narrowest trace possible that is consistent with achieving his/her temperature constraints. All of the above suggests that can't be achieved with charts, equations, or formulas. Designers now need thermal simulation techniques to achieve their goals. And there is a precedent for this. In the 1990s many designers began to worry about controlled impedance traces.[4] At first we had equations that we could use, but in today's world, equations are not good enough. In the industry parlance, field effect solutions are required. Thermal simulation models, like TRM, employ the same quantitative techniques that field effect solutions employ.

That is not to say that there aren't some general conclusions we can derive. For example, looking at Figure 7.17, a 10.7% increase in resistivity leads to an 8.1% increase in trace temperature. It may pay to look at the copper characteristics of the materials used by the material supplier and the board fabricator. Similarly, if the thermal conductivity coefficient can vary by, say, 0.3 between material choices, that may lead to a temperature reduction of as much as 10°C.[5]

7.5.1 Call to Action

It is apparent from the above analysis that there are two important areas where additional research is necessary. First, we do not know enough about how the HTC convection values change with temperature and with conditions. A trace represents a small hot spot on an otherwise moderately uniform plane. We need to come up with ways to better estimate this effect.

Second, the industry needs to do a better job in providing thermal conductivity information about the materials that are supplied. But further, the authors hypothesize that (a) since the board material conducts heat in the in-plane direction more efficiently than it does in the through-plane direction, (b) therefore there should be a correlation between the ratio of resin to fiberglass and the trace temperature.

In addition, are there compounds that can be (economically) added to board materials (dielectrics) that can increase their thermal conductivity? For example, are there changes to the resin formulations that might accomplish this? Such approaches might improve the heat-spreading capabilities of board materials almost to the point approximating the effects of underlying planes.

End Notes

1. In fact, this would be a terrific avenue for further research.
2. The authors discuss how to measure the resistivity of copper traces and the difference between ED and rolled copper in Appendix B.
3. For comparison, the thermal conductivity coefficient for copper is 385 (http://hyperphysics.phy-astr.gsu.edu/hbase/tables/thrcn.html).
4. Controlled impedance traces are those designed to look like transmission lines, so those lines can be terminated in their characteristic impedance, thereby reducing signal reflections back on the traces.
5. This would apply for trace temperatures in the range of 50 degrees or so. For higher trace temperatures, the difference would be even greater.

CHAPTER

8

Contents

Via Temperatures

8.1 Bottom Line

- Via temperatures are *not* related to the current through the via. Via temperatures are determined by the temperature of the parent trace.
- As long as the parent trace is sized correctly, a via of any size cannot be much warmer than the trace.
- Voltage drops across vias are negligible and can be ignored.
- Thermal vias are not as effective as the planes or heat sinks they connect to.

8.2 Background Information

Through the years there has been very little written about the current carrying capacity of a via. This is, we believe, because there has been no practical way to measure or to predict the temperature of a via. A thermal simulation model gives us a unique way to look inside a board and inside a via to predict what is going on. When we did that with TRM, we were amazed at what we found. But after a little thought, what we found made perfect intuitive sense. It is just that we had been viewing the problem incorrectly for all these years.

A via typically has a drill diameter and a copper conducting wall. The thickness of the wall is formed during a copper plating process. This gives the via a conducting cross-sectional area. Consider the situation where a via connects a plane or trace on one layer to a plane or trace on another layer. If that via will carry a significant current (i.e., a current high enough to heat a trace to some degree), a relevant question is how big do we need to make the via's (conducting) cross-sectional area in order to safely handle that current? Until now there is been very little practical guidance on that question. But a thermal simulation model provides us with a tool to begin to explore that question.

Board designers typically approach this question using one of three strategies:

1. Simply don't allow vias that carry higher levels of current. Route all the conductors on the same trace layer.
2. Size the via to handle the current using the IPC-2152 guidelines.
3. Use a standard via known to be able to carry a certain amount of current and use multiple vias in parallel, as many as needed for the total current carried by the trace.

IPC-2152 explicitly endorses 2 and/or 3 on page 26:

> *The cross-sectional area of a via should have at least the same cross-sectional area as the conductor or be larger than the conductor coming into it. If the via has less cross-sectional area than the conductor, then multiple vias can be used to maintain the same cross-sectional area as the conductor.*

Anecdotally, this must be good advice because most designers have never had a via fail for purely thermal reasons (unless there was a fabrication or alignment issue). What we have been doing must be right, because it works so well! In fact, it turns out this guidance has been *extremely* conservative.

But there is an implicit assumption in this guidance that sometimes poses a problem. That is, it is assumed we know the plating thickness of the via. We typically assume (*hope* might be a better word) that the via wall thickness is the same as the plating thickness, which is typically 0.5- or 1.0-oz copper (0.65 to 1.3 mils). But the plating is often not uniform on the walls of the via. This is a separate problem that we are not prepared to address in this book.

8.3 Thermal Simulation

In Chapter 6 we described the basics behind a thermal simulation model. Think of the board being modeled as consisting of a large number of small cubes. Each cube is small enough so that the parameters don't change much across any of its surfaces. Therefore, the computer model solves for the boundary conditions of each cube, and then starts assembling a complete model.

The sides of the thermal cubes must be smaller than the smallest dimension (in the *x,y* planes) on the board being modeled.[1] For a via simulation, this is the thickness of the conducting surface on the wall of the via. This is a much smaller dimension than most thermal simulation models deal with, and indeed smaller than any models developed so far in this book. The problem is that such small cubes means that there are *lots* of cubes making up the model. That, in turn, means the matrices are very large, and that means the central processing unit (CPU) load on the computer is very large (meaning long processing times).

The practical response to this issue is to model a board that is smaller than what a conventional board might look like. This will (probably) result in higher simulated temperatures than the actual temperatures, but not too much higher. More importantly, any *relationships* between and among various modeled traces will be preserved.

8.3.1 Simulation Strategy

So we are going to proceed with the following set of assumptions and strategies:

- We will assume a standard via with a 10-mil (0.26-mm) drill diameter (radius = 5 mil) and a 0.030-mm uniform plating thickness. This is slightly less than a 1.0-oz equivalent. The effective conducting cross-sectional area of this via is $\pi(r^2 - (r\text{-}th)^2)$, where r is the via radius and th is the plating thickness. Choosing these millimeter equivalents (which are slightly rounded) results in a via conducting cross-sectional area of .0217 mm^2. This is approximately equivalent to a 1-oz, 26-mil (.66-mm) wide trace.
- The pad area of a via (top and bottom layers) doesn't matter because we are going to be dealing with traces that are at least 16 mils (0.41 mm) wide.
- We will be dealing primarily with a standard board thickness of 63 mils (1.6 mm).

- We will vary the trace width and current and compare the temperature of the simulated trace with the simulated via temperature. In this way, we will see if the via gets hotter than the trace, stays the same temperature as the trace, or runs cooler than the trace.
- We assume that at low currents via temperature is not a problem. Since we are primarily concerned with high-temperature effects, we will concentrate on what happens at higher temperatures and currents.

8.3.2 Board Model

Our thermal model is a pair of 60-mm long traces on either side of a board connected by a 0.26-mm diameter, 0.03-mm plated via. The board is a few millimeters wider than the trace. The dielectric layer between the trace layers is 1.6 mm FR4. A model of a thermocouple (TC in Figure 8.1) is placed near the midpoint of the top trace. This will measure the temperature at that point during the simulations. The thermocouple will be an approximate measure of what the temperature of the trace would be without a via. A diagram of this model is shown in Figure 8.1.

8.3.3 First Simulation

Our first series of simulations will be run with this model in five widths: 0.40, 0.66, 1.2, 1.8, and 2.5 mm. (This approximately corresponds to widths of 16, 26, 47, 71, and 100 mil, respectively.) The results of the first simulation, 0.66 mm, are shown in Table 8.1. This is the simulation where the via and trace conducting areas are approximately equal. The values shown are the actual temperatures (*not* the temperature *change*) at the thermocouple

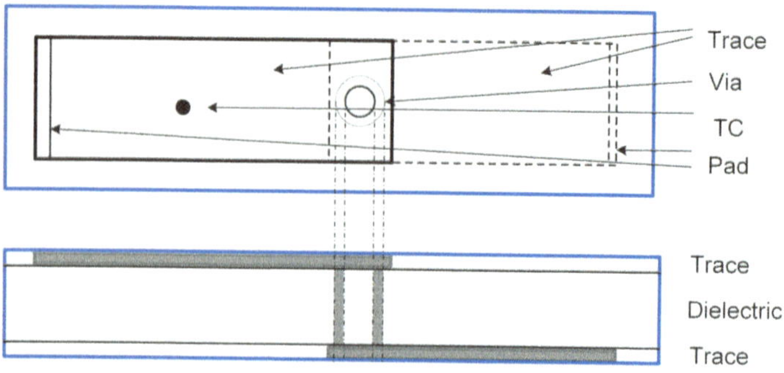

Figure 8.1 Thermal model of trace with via (not to scale).

Table 8.1
Temperatures in the 0.66-mm Simulation

Current (A)	Thermocouple (°C)	Via Top (°C)	Via Midpoint (°C)
3.0	86.9	81.4	80.9
4.0	166.7	154.1	152.9

(TC), the maximum temperature point of the upper trace, the top of the via, and the midpoint of the via.

One of the first questions to ask about the simulation temperatures is "compared to what?" We also modeled each configuration as a single trace on a single layer with no via under the same conditions. That way we can compare the via temperatures with what the trace temperature would be without the via. The thermocouple is placed so that it will provide what the trace temperature would be without the via.

In this first simulation, the via conducting cross-sectional area and the trace conducting cross-sectional area are approximately the same. Under the assumption that the cross-sectional area of the via determines the via temperature, the via midpoint temperature should equal the same as the TC (which equals the temperature of a trace without a via). We can see from Table 8.1 that *this is not true*. At 3.0 amps the via runs approximately 6°C *cooler* than the trace and at 4.0 amps it runs approximately 12°C to 13°C *cooler*. The thermal profile of the top trace layer at 3.0 amps is shown in Figure 8.2. It is apparent that the trace begins to *cool* as it approaches the via in the center.

8.3.4 Additional Simulations

We ran additional simulations at four other widths. The results for all five simulations are shown in Table 8.2 and graphed in Figure 8.3. The graph sets from left to right are for 0.40-, 0.66-, 1.2-, 1.8 , and 2.5-mm traces, respectively (approximately 16, 26, 47, 71, and 100 mil).

It is apparent from the data that

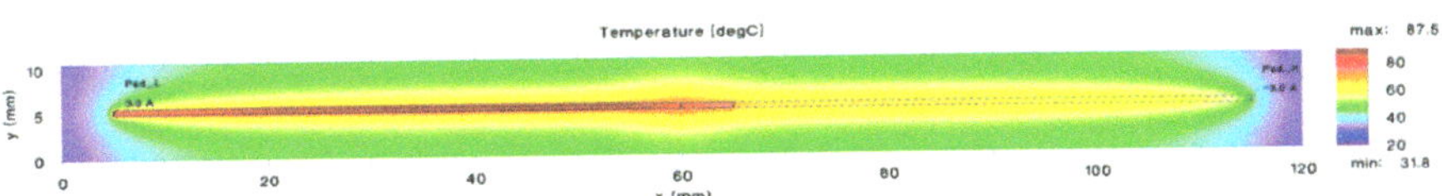

Figure 8.2 Thermal profile of the top trace layer of the first simulation at 3 amps.

Table 8.2
Temperatures in the Simulation at Various Points and Currents

Width (mm)	Current (A)	Thermocouple (°C)	Via Top (°C)	Via Mid-point (°C)
0.40	2.0	91.4	83.2	82.4
0.40	2.3	123.2	111.1	109.9
0.66	3.0	86.9	81.4	80.9
0.66	4.0	166.7	154.1	152.9
1.2	3.0	47.9	48.6	48.7
1.2	4.0	74.1	75.7	75.8
1.2	5.0	115.5	118.8	118.8
1.8	5.0	70.8	76.8	76.8
1.8	6.0	100.2	110.2	110.2
2.5	6.0	68.1	77.6	77.6
2.5	7.0	90.3	104.9	104.9

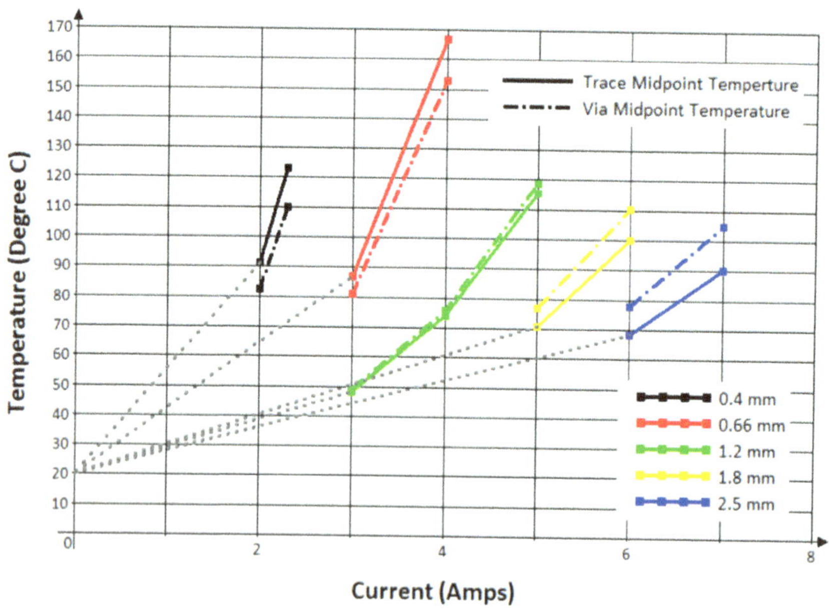

Figure 8.3 Via and trace temperature versus current for 1.0-oz traces of varying width.

1. If the conducting area of the trace is smaller than that of the via, the trace is hotter than the midpoint of the via.
2. If the conducting area of the trace is larger than that of the via, the trace is cooler than the midpoint of the via.

3. Therefore, there must be a trace conducting area where the trace temperature and the via midpoint temperatures are equal. The simulations suggest that this occurs when the trace conducting area is approximately 1.5 to 2.0 times the via conducting area.

Looking at the 0.66-mm simulation, the reason that the via runs cooler than the trace (for the same conducting areas) has not been obvious up until now. One of the surprises in IPC-2152 is that internal traces run somewhat cooler than do external traces of the same dimension (refer back to Sections 5.4.3 and 7.4). That is because, it turns out that the heat spreading through the dielectric supports more cooling than does convection and radiation into the air. The via looks very similar to an internal trace surrounded by a dielectric. That is why it runs cooler than an equivalent trace on the top layer. Furthermore, the copper is an extremely good conductor of heat. The via and the trace have the same cross-sectional area, and therefore the same resistance per-unit-length. Therefore, we can expect them both to heat approximately equally through the I^2R heating mechanism. Since the via runs cooler than the trace, what is happening is that *the via is helping to cool the trace.*

As the trace width increases, and as the current increases, the via does start running a little hotter than the trace, but only by a few degrees. As the trace width gets wider, and therefore can handle the thermal heating better than the via can, the trace starts cooling the via (because of the high thermal conductivity between the via and the trace). In effect, the trace becomes a heat sink for the via. For example, look at the comparison in Table 8.3, a subset of the data in Table 8.2.

The via cross-sectional area is about the same as that for a 0.66-mm trace. Under the old guidelines, we would expect the via temperature to increase with current. But the via midpoint temperature stays approximately the same as we increase the current and the trace width at the same time. At 6.0 amps the via is running *cooler* with a 2.5-mm trace than it does at 2.0 amps with a 0.40-mm trace!

Table 8.3
Subset of Data in Table 8.2

Width (mm)	Current A	Thermocouple (°C)	Via Top (°C)	Via Midpoint (°C)
0.40	2.0	91.4	83.2	82.4
0.66	3.0	86.9	81.4	80.9
1.2	4.0	74.1	75.7	75.8
1.8	5.0	70.8	76.8	76.8
2.5	6.0	68.1	77.6	77.6

The comparison is even more dramatic if we simulate a 2.0-oz trace. We simulated a 2.5-mm, 2.0-oz trace carrying 7 and 10 amps with only a single via. The results are shown in Table 8.4. Note that with 10 amps the via midpoint temperature is about 118°C, almost 25 degrees hotter than the trace TC. But consider this: a single 0.66-mm trace (with the same cross-sectional area as this via) carrying 10 amps would reach the melting point of copper (1083°C) in approximately 7.5 seconds![2] The via only reaches 118 because of the thermal coupling (heat sinking) to the 2-oz trace.

All the above simulations used enough current to raise the temperatures (not the change in temperatures) to a pretty high level. This was to exaggerate the results. Most designers would not design to these levels. Typically, designers target for only a 20°C to 40°C temperature rise maximum. But what if we use our standard via (designed for a 1-oz, .66-mm trace) with a 2.0-oz trace with only enough current to raise the temperature by 20 or 30?

We ran one final simulation of a 2.0-oz, 2.5-mm wide trace. The results are shown in Table 8.5. At 7.0 amps, the via midpoint temperature is almost 41 above ambient (of 20°C; see Table 8.4), while the trace is about 31 degrees above ambient. At 5.0 amps (Table 8.5), the trace is only 17.1°C above ambient and the via just slightly higher at 21.1(°C) above ambient. This is in spite of the fact that the same via with only 4.0 amps is 145(°C) above ambient (see Table 8.2) in a 0.66-mm trace with the same conducting area as the via!

8.3.5 Two Vias

Even though a single via can carry significantly more current than expected, as a result of the thermal coupling to the trace, the capacity is of course greater if we use two vias. Table 8.6 provides the results of simulations of

Table 8.4
2.0-oz Trace with a Single, Small Via

Current (A)	Thermocouple (°C)	Via Top (°C)	Via Midpoint (°C)
7	51.3	59.0	60.8
10	93.5	113.2	117.9

Table 8.5
Temperature for 2.5-mm Trace, Single Via

Current (A)	Thermocouple (°C)	Via Top (°C)	Via Midpoint (°C)
5.0	37.1	40.3	41.1

Table 8.6
Temperatures for 2.5-mm Trace at Various Points and Currents, Two Vias

Weight (oz)	Width (mm)	Current (A)	Thermocouple (°C)	Via Top (°C)	Via Midpoint (°C)
1.0	2.5	6.0	67.1	69.5	69.5
1.0	2.5	7.0	88.6	92.3	92.4
2.0	2.5	6.0	41.3	43.3	43.5
2.0	2.5	7.0	49.9	52.7	53.0
2.0	2.5	10	89.6	96.5	97.3

two different 2.5-mm (100-mil) traces, each with two vias. One is a 1.0-oz trace and the other is a 2.0-oz trace. The temperatures at the two vias on each trace are virtually identical, confirming the fact that the current is splitting equally between the two vias. The graphical results are shown in Figure 8.4. The black lines are the via midpoint temperatures and the red lines are the trace TC (midpoint) temperatures.

Note that even at 10 amps through a 2.0-oz trace, the small vias are less than 10 degrees warmer than the trace.

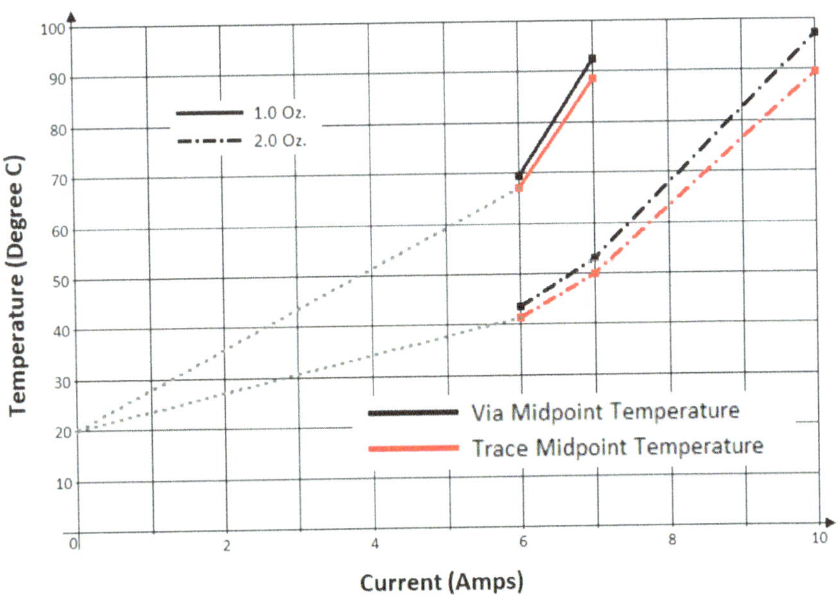

Figure 8.4 Trace and via temperatures for a 2.5-mm (100-mil) wide trace with two vias.

8.3.6 Conclusion

The inescapable conclusion from the above is that the conventional wisdom has been wrong! The current does not determine the temperature of the via—the associated traces do. As long as the traces are sized to properly handle the current, even a single, ordinary via is adequate to transition between trace layers. And the reason is because of the thermal conductivity of the copper between the via and the trace. The traces act as a heat sink for the vias

If there was ever a conclusion that cried out *"SHOW ME!"* this is it!

8.4 Experimental Verification

To test the results of the previous section, the first thing we need is a test board. Prototron Circuits (Tucson, AZ) was very generous in providing test boards for this purpose. The relevant portion of the test board is shown in Figure 8.5. The board is approximately 60-mils thick FR4.The board contains 0.5 oz copper nominally plated with 1.0 oz additional copper. Two traces were compared, one nominally 27 mils wide, the other 200 mils wide. Each trace is 6.0 inches in length, one-half on the top layer and one-half on the bottom layer. Each trace has a single 10-mil diameter plated via connecting top to bottom. It is important to note that the via structure is identical in each trace. The board was microsectioned after testing to measure the actual dimensions for the simulation.[3]

The 10-mil diameter via plated to 1.0 oz has roughly the same conducting cross-sectional area as the 27-mil trace. That is why the 27-mil wide trace was chosen for this study.

8.4.1 Simulation

Before we tested the actual board, we felt it was important to replicate the simulation results on the actual board used for testing. For the reasons

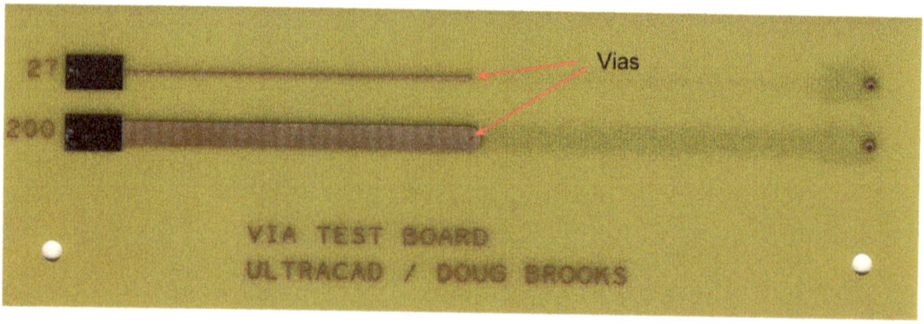

Figure 8.5 Relevant portion of via test board.

described above (Section 8.3), the simulation model has to be smaller than the actual board, leading to the possibility of slightly higher simulation results than actually measured. It turned out there was one other measurement that was necessary, determining the *actual* thermal conductivity of the board material. C-Therm Technologies[4] graciously measured the thermal conductivity of the board material to facilitate the simulation. They measured the thermal conductivities as (W/mK):

In-plane: 0.679

Through-plane: 0.512

It turned out these thermal conductivity values are significantly higher than those modeled in Section 8.3, which resulted in lower overall simulated temperatures. Also, in Section 8.3 our modeled trace and the via wall had the same thickness (1.0 oz). The actual test board has plated copper over 0.5-oz foil, so the nominal trace thickness will be 0.5 oz thicker than the via wall. This will also lead to lower overall experimental temperatures.

8.4.2 Simulation Results

Table 8.7 shows the result of the simulation. They are consistent with the type of results presented in Section 8.3. For a small trace that is approximately the same size as the via, the via is cooler than the trace because the via cools more effectively into the dielectric layer of the board than the trace does. For the larger trace carrying more current, the heat generated in the via is conducted away from the via onto the trace, resulting in the via being a little hotter (but not a *lot* hotter) than the trace.

Figures 8.6 and 8.7 are the simulated thermal profiles of the 200-mil trace carrying 8.55 amps and the 27-mil trace carrying 6.65 amps, respectively. Note that the thermal profile cools a little near the via for the 27-mil trace. The via is the hottest point on the 200-mil trace, but only 12% hotter than the trace itself.

Table 8.7
Simulation Results for Test Board

Trace Width (mils)	Current (A)	Trace Temperature (°C)	Via Temperature (°C)	Via T/ Trace T
27	4.75	72.8	70.1	96.3
27	6.65	114.2	108.2	94.7
200	4.75	30.8	31.8	103.2
200	8.55	44.8	48.1	107.4

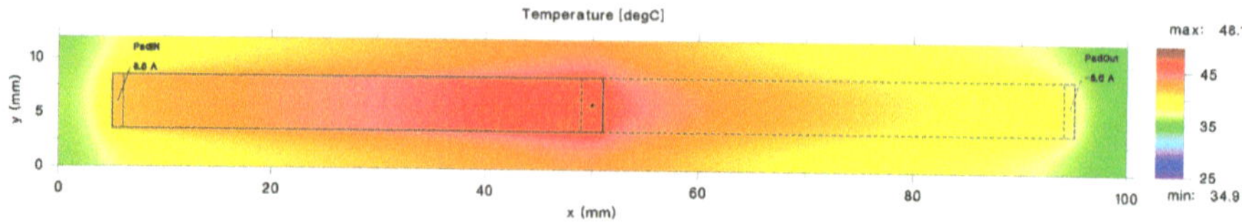

Figure 8.6 Thermal profile of 200-mil trace carrying 8.55 amps.

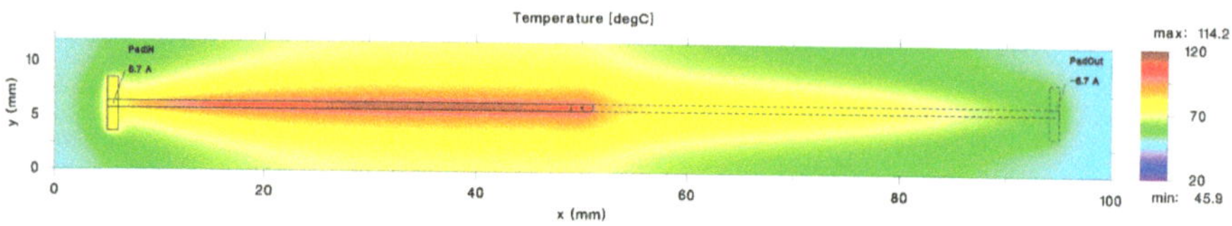

Figure 8.7 Thermal profile of 27-mil trace carrying 6.65 amps.

8.5 Experimental Results

The test procedure is pretty straightforward. A constant current is applied to the trace. The current level is checked and confirmed by a second digital meter. The temperature is measured with a thermocouple probe whose tip diameter is spec'd at 30 Ga (approximately 10 mil diameter). The probe response time is approximately 1 second. The probe was calibrated to ice water and to boiling water at an elevation of 360 feet and found to be well calibrated. Various tests were run to verify that the probe does not affect the temperature of a trace being measured.

For each test, a constant current is applied to the trace until the temperature stabilizes (approximately 6 minutes). Then the temperature is measured and recorded. After the measurements were taken, the board was returned to Prototron for microsectioning to confirm all dimensions.

8.5.1 Measured Results

The measured test results are provided in Table 8.8.

There are two especially important observations that should be made here:

1. First, a 6.6-amp current through the 27-mil wide trace results in a via temperature of 109(°C)while a *higher* current of 8.6 amps results in a larger (200-mil wide) board results in a *much* lower via temperature of only 44.5°C. This confirms that it is the trace that is controlling the via temperature.

Table 8.8
Measured Test Results

Trace Width (mils)	Current (A)	Trace Temperature (°C)	Via Temperature (°C)
27	4.75	66	64.5
27	6.65	114	109
200	4.75	30.5	31.5
200	8.55	40.5	44.5

2. The measured data (Table 8.8) is very close to the simulated data (Table 8.7). This gives us confidence that the simulation approach is a viable approach for predicting temperatures in complex situations.

Figures 8.8 and 8.9 show thermographs of the 27- and 200-mil traces carrying 6.65 and 8.55 amps, respectively. Note the similarity to the simulated thermal profiles in Figures 8.6 and 8.7. In particular, Figure 8.8 shows the hottest part of the trace well to the left of the via, consistent with the above results. Figure 8.9 shows the hot spot at the via, but it is only a few degrees hotter than the rest of the trace, and it is much cooler than the via in Figure 8.8, even though it is carrying more current.

8.5.2 Conclusion

The results of this experimental evaluation are consistent with, and would seem to confirm, the results predicted in Section 8.3. It is not the current that determines the temperature of a via, it is the temperature of the associated traces. If the traces are sized correctly for the current level, *much* smaller and fewer vias are required to transition between trace layers than

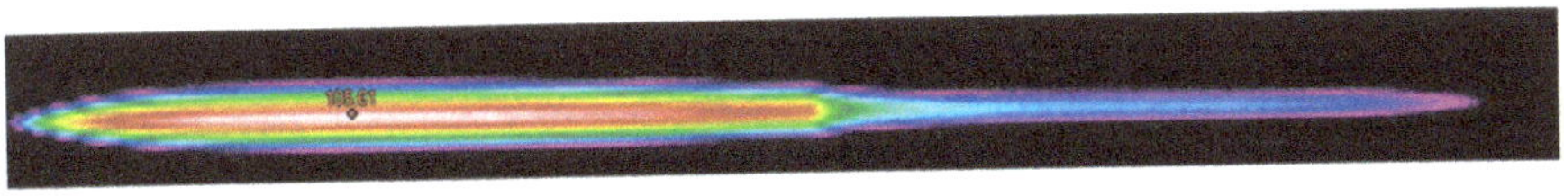

Figure 8.8 27-mil trace carrying 6.65 amps.

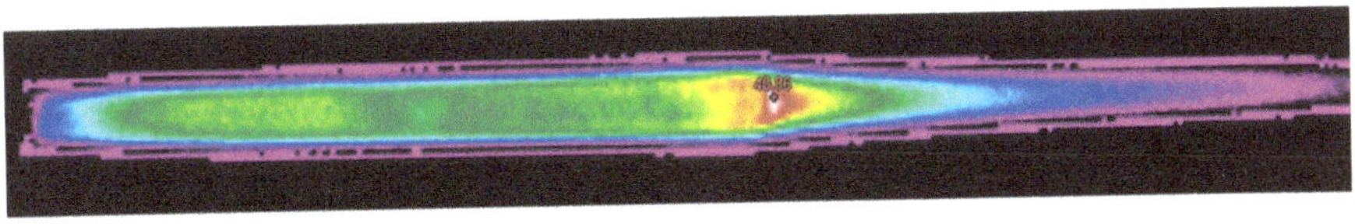

Figure 8.9 200-mil trace carrying 8.55 amps.

has been previously believed. This means designers can have much greater flexibility in freeing up routing channels underneath the traces on their boards.

8.6 Voltage Drop Across Trace and Via

Designers and engineers are sometimes concerned about the voltage drop across a trace or a via. TRM simulations allow us to analyze this fairly easily. We will use the 26-mil (.66-mm) trace we modeled with 3.0 amps of current in Section 8.3.3.

We modeled this trace, as well as a trace with identical dimension but no via, using the latter as a control. TRM allows us to look at the voltage gradient at any point along the trace (just as it allows us to look at the thermal gradients, as we have seen, and at the current gradients, as we will see in the next chapter). The voltage drops we derived using TRM were

.66-mm wide trace without via: 0.34 volts

.66-mm wide trace with via: 0.32 volts

Volts across via: 0.00 volts

Interestingly, TRM measured *no* voltage across the via and *less* voltage across the trace *with* a via than the trace *without* the via. The reason the voltage across the trace with the via is smaller is because the trace with the via is cooler than the trace without the via (see Table 8.1). That is because the via is helping to cool the trace. The reason there is no (actually a very small) voltage across the via is because the via is so short.

We can check some of these figures with the following equations:

$$R = (\rho / A) * L \tag{8.1}$$

$$R(T) = R(T_o) * (1 + \alpha_o * (T - T_o)) \tag{8.2}$$

First, let our parameters be

$\rho = 1.7\ \mu$ohm-cm

$\alpha = 0.004$

$T_o = 20°C$

$T = 86°C$

$L = 115$ mm (4.5 inches)

W = .66 mm (26 mils)

Th = 1.0 oz

These parameters lead to an initial resistance of 0.0897 ohms at 20°C and a final resistance of 0.11339 ohms at 86°C. Then, using Ohm's law, we see that the calculated voltage drop would be 0.11339 ohms * 3.0 amps = 0.34 volts, just as TRM predicts.

Then, the voltage drop across the entire trace, 0.34 volts, occurs across 115 mm. But our via length is only 1.6 mm (63 mils), the thickness of the board, and so we can approximate the voltage drop across the via as

$$V\text{via} = 0.34 * 1.6/115 = 0.0047 \text{ volts} \tag{8.3}$$

That is, the voltage drop across the via is negligible compared to the voltage drop across the entire trace.

In Section 8.3.4 we also modeled a 100-mil (2.5-mm) wide trace with 7 amps, reaching a trace temperature of 90°C, slightly higher than the 26-mil wide trace with 3.0 amps. The resistance and voltage drop for that trace (without a via) would be (using(8.1) and (8.2)):

R at 20°C 0.0237 ohms

R at 90°C 0.0303

V = .0303 * 7 amps = 0.212 volts

Assuming the via remains the same resistance (in this illustration it would be slightly higher), we can again approximate the voltage drop across the via by using these proportions:

$$\text{Via voltage (at 7 amps)} = .0047 * 7 / 3 = 0.011 \text{ volts}$$

This is still substantially below other voltages along the trace.

8.6.1 Summary

The voltage drop across a trace with a typical via is lower than we might expect, based on the maximum temperature of the trace, because the via helps cool the trace. And since the length of a via is generally much less than the length of the trace, the additional resistance of the via contributes a negligible increase in voltage drop. In general, from a practical standpoint, we can consider a voltage drop across a via to be insignificant compared to the voltage drop across the traces.

8.7 Thermal Vias

Designers sometimes drop vias between a plane or pad to conduct heat away to another plane or heat sink. This is thought to be an effective way to cool a component that generates a lot of heat at a concentrated point on a board.

TRM offers an excellent way to test this thought. Our simulation model will be a simple board measuring 60 × 120 mm. It will be 1.5-mm (60-mil) thick FR4. There will be a 20- × 20-mm pad at the center of the board representing a hot component. The pad will be 33 μm thick, approximately equivalent to 1.0 oz copper. We will apply 40 amps across the pad to generate the heat. For a simple, isolated pad, this will raise the temperature to 82.1°C, 62.1°C above the ambient temperature. Our simulated thermal via will be 0.5 mm (20 mil) in diameter. We will assume it is solid copper throughout. We will add one (centered) or five (equally spaced) thermal vias to our simulation.

Now here is an important point. A thermal via presumably has to go somewhere. We will simulate three possible conditions:[5]

1. There is a small plane on the opposite side of the board, the same size as our heated pad.
2. There is a large plane on the opposite side of the board, much larger than our heated pad.
3. There is a special 20°C heat sink on the opposite side of the board the same size as our heated pad. It is special in the sense that it does not change temperature (e.g., it is cooled by flowing water).

Table 8.9 and Figure 8.10 shows the results of the simulations. The temperatures shown in the figure are the *maximum changes* in temperature of the pad under the conditions simulated. The small plane does not help much in cooling the pad. Each thermal via helps cool the pad another increment. But the larger plane (by itself) substantially cools the pad above it (see Figure 8.11). As we saw in Section 7.2.6, an underlying plane can cool a trace; here it cools a component pad. The underlying heat sink is even more effective in cooling the pad.

In each of the cases, the addition of thermal vias offers some additional cooling, but in the case of the larger plane and the heat sink their marginal contribution is much smaller. Most of the pad cooling (at number of thermal vias = 0) has been accomplished by the (necessary) plane or heat sink. We can understand why planes/heat sinks are more effective then thermal vias by looking at the basic conductive heat transfer formula in (8.4):

Table 8.9
Simulation Results for the Maximum Temperatures

Simulation	Pad Tempurature (°C)	Pad ΔT (°C)	Plane/Heat Sink Temperature (°C)	Plane/Heat Sink, ΔT (°C)
No plane or via	82.1	62.1	na	na
Small plane only	79.6	59.6	73.5	53.5
Small plane + 1 via	78.7	58.7	76.0	56.0
Small plane + 5 vias	67.5	47.5	66.0	46.0
Large Plane only	43.0	23.0	35.8	15.8
Large plane + 1 via	42.1	22.1	38.7	18.7
Large plane + 5 vias	36.4	16.4	34.7	14.7
Heat sink only	29.5	9.5	20.0	0.0
Heat sink + 1 via	28.1	8.1	20.0	0.0
Heat sink + 5 vias	24.2	4.2	20.0	0.0
Special via	42.6	22.6	na	na

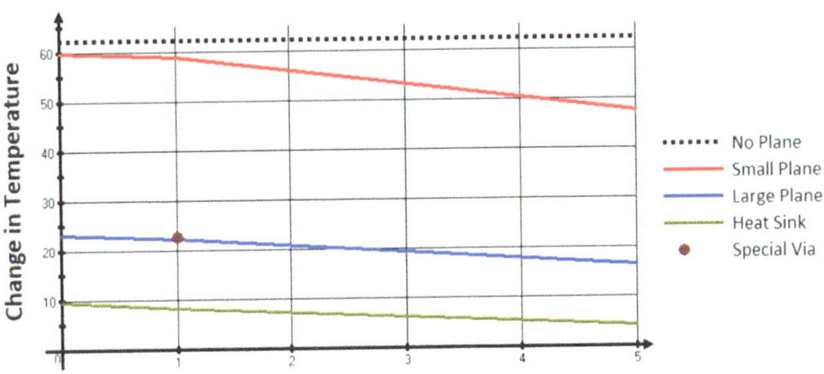

Figure 8.10 Thermal via simulation results.

$$Q/t = kA(\Delta T)/d \tag{8.4}$$

where

Q/t = rate of heat transfer (watts or joule/sec)

k = thermal conductivity coefficient (W/mK)
about 0.5 for FR4

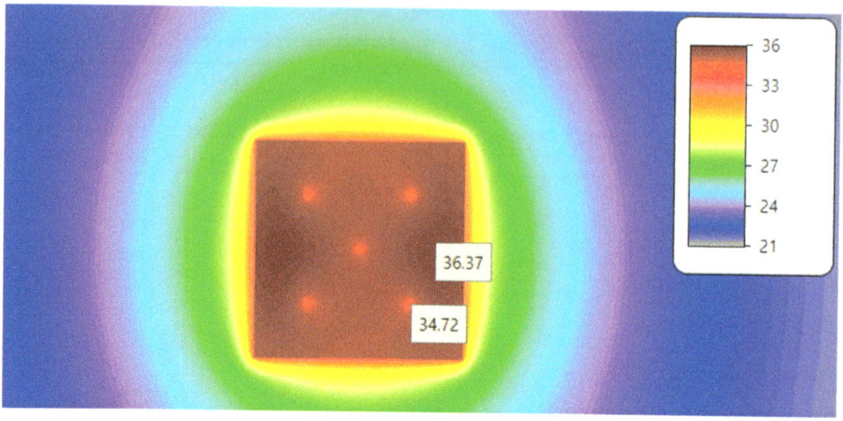

Figure 8.11 Thermal image of pad with five thermal vias over the large plane.

about 385 for copper

ΔT = change in temperature (°C = °K)

A = overlapping area

= 400 mm^2 for the pad

= πr^2 = (3.14)(0.25^2) = 0.196 mm^2 for the thermal vias

d = distance between the pad and plane/heatsink = 1.5 mm

A/d = 400/1.5 = 267 mm for the pad

= .196/1.5 = 0.131 mm for the thermal via

The heat transfer rates for the pad and the thermal vias are different, and since the temperatures change with time, Q/t is a point (in time) concept (see Section 7.2.2). We can compare the transfer rates using (8.5) (the ds cancel, as do the area units):

$$(Q/t)_p/(Q/t)_{tv} = [kA(\Delta T)]_p/[kA(\Delta T)]_{tv} \tag{8.5}$$

where the subscripts p and tv mean pad and thermal via, respectively.

One can argue whether the two ΔTs (numerator and denominator) are equal and cancel. One argument is that they are different, except that ignores the heat flow path from the hottest to the coolest points on the respective planes. One can argue that the appropriate measure would be from the average temperatures on the two surfaces, except that is (a) difficult to calculate and (b) difficult to defend. We are going to let them cancel because, in the big picture, it doesn't make a big difference in the final implications. So (8.6) compares the heat flow rates of the planes to the thermal via(s) (again, length units cancel):

$$(Q/t)_p/(Q/t)_{tv} = (kA)_p/(kA)_{tv} = (0.5)(400)/(385)(0.196) = 2.65 \quad (8.6)$$

The overlapping planes (in this example) are 2.65 times more effective in conducting heat than is a thermal via. But in addition, if we add an additional thermal via, the ΔT has already been dramatically reduced by the plane itself. That is why we see in Figure 8.10 that the marginal improvement of adding thermal vias in the large plane or heat sink cases is nominal. The big benefit has already been achieved by the plane or heat sink in the first place. Only in the case of a very small plane is the addition of more thermal vias significantly beneficial.

8.7.1 Special Via

We added one more simulation to the above for interest. What if we could add a *special* via to the pad? This via would have the same diameter as the other thermal vias but would perform as a spot heat sink, holding the temperature *at that spot* to 20°C. It is hard to imagine how we would do that, but if we could, what, then, would be the impact on the rest of the pad?

The result is shown in Figure 8.10 by the point labeled "special via." The temperature at the center of the pad is drawn down to 20°C. Then the temperature increases as we move away from that point. The maximum pad temperature would be 42.6°C, an increase of 22.6°C above the center point. This has almost exactly the same impact as the large plane and a single (regular) thermal via.

8.7.2 Conclusion

Once we recognize that thermal vias, almost by definition, require another surface to connect to, we conclude that the thermal benefits are provided primarily by thermal conduction through the dielectric to the other surface rather than the thermal vias themselves. Only in the cases where the secondary surface area is small do thermal vias have more significant effects.

End Notes

1. This is related to the thermal pixel size referenced in Section 6.2.
2. See Chapter 12.

3. The trace widths were essentially as designed. The thicknesses were (top layer) 2.1 mil and (bottom layer) 2.9 mil.
4. See their measuring process described in Appendix A.
5. There are an almost infinite number of design variations that might apply to a thermal via situation. These simulations are just three illustrations.

CHAPTER

9

Contents

Current Densities in Vias

9.1 Bottom Line

- Current tends to follow the shortest path.
- Current divides equally among vias for traces in a straight line.
- Current density is higher at the top of some vias than current flow would indicate. There is some double counting of current around the top of some vias.

9.2 Background

While we were investigating the relationship between current and via temperature, we noticed a very interesting pattern in the current density at various points within a via. We ran a separate series of simulations designed to look more closely at current density. (Note, however, that in Chapter 14 we point out that current density has little relevance in trace current/temperature analyses.)

We reproduce and discuss current densities here, not because they are particularly relevant to any other discussions in this book, but because they may be instructive to the reader as to how current flows through a circuit. We all learned long ago that current follows the (shortest) path of least resistance (actually least inductance, but we are talking about DC here). In the figures in this chapter we can actually visualize how that is happening as current wraps around the wall of a via. We will see a similar phenomenon in Section 13.4 as we look at the current density around a right-angle corner.

9.3 Single Via

Our first simulation was of a simple trace on the top layer of a board connecting to a trace on the bottom layer by means of a through-hole via. The trace was 0.33 mm (13 mil) wide and 34 μm (1.0 oz) thick. The board dielectric was FR4, 1.61 mm (63 mil) thick. The board itself was a few millimeters wider than the trace. For simulation purposes, the dielectric was modeled as a set of seven individual layers, each 23 μm thick. Using multiple layers in the simulation allows better precision in the model, as well as allowing us to look at the current densities at various points along the via wall. The via was .26 mm (10 mil) wide with a wall thickness of .03 mm (slightly less than a 1.0-oz equivalent). This results in a via conducting area of .021677 mm^2 (see Figure 9.1).

The current density patterns of the first two layers (trace layer and Core 1) are shown in Figure 9.2. Section F.1 in Appendix F offers a detailed explanation of how to interpret these figures, and Section F.2 provides the full set of patterns through Core 4. We encourage the reader to read Section F.1 before proceeding further in this chapter. We will look at the current density pattern for the Core 1 layer first, then come back to the top (trace) layer.

Figure 9.3 shows the current density pattern of the Core 1 layer in more detail. The current density (in A/mm^2) varies from 223.7 on the near (leading) side to 140.4 on the far side of the via wall. If we total these four

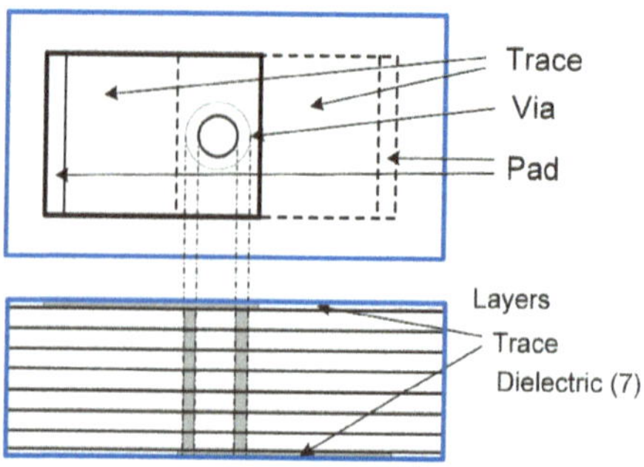

Figure 9.1 Simulation model of a single via (not to scale).

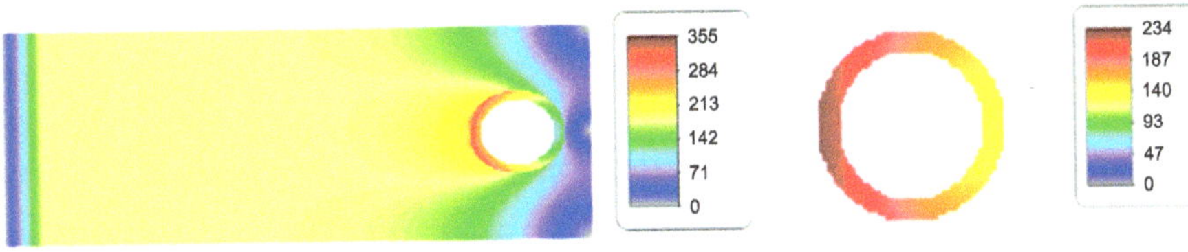

Figure 9.2 Current densities (amp/mm^2) on the top, trace, layer (left), and the first core layer (right).

Position	Density
1	223.7
2	179.5
3	140.4
4	180.2
Total	723.8
average	181.0
x area	3.922 A

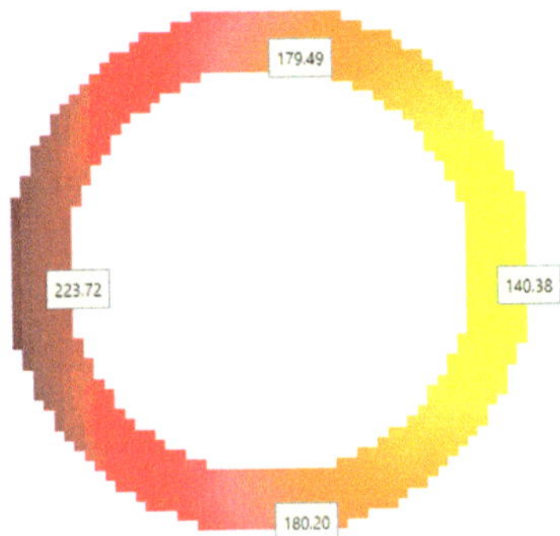

Figure 9.3 Current densities, Core 1.

density samples we get 723.8, which, if divided by the four measurements, results in an average density of 181.0 A/mm^2. If we then multiply this average density by the area of the via wall (.021677 mm^2) we get 3.922 amps.

We would expect that product to be the 4.0 amps we applied to the trace in the simulation. This result, which is typical of all the simulations to follow, represents about a 2% difference from what is expected. This is because we are approximating a cylindrical (via) wall with a number of squares in our computer model (see Section 6.3 and especially Figure 6.1, and also Section F.1.1 in Appendix F). The dimension of each square is a thermal pixel (see Section 6.4).

The square approximations of each core are visible in each core image. The smaller the thermal pixel, the better the approximation. But the smaller the thermal pixel, the more pixels there are in the simulation and the greater the calculational load in the computer CPU.

As can be seen from Section F.2, as we go from Core layer 1 to Core layer 4, current densities decrease at the front of the via and increase at the rear of the via. At Core layer 4, the midpoint of the via, the current density is uniform around the via wall (at 181.3 A/mm^2).

TRM simulations are steady state (think DC) in nature. We normally think of current density being uniform in a conductor in steady state.[1] So why are there gradients in the via wall? The answer relates to the length of the current path. The *shortest* current path starts at the front (leading edge) of the via, spirals around each side of the via as you progress downward through the via, and exits the via at the leading edge of the via *with respect to the bottom trace* (which is at the far side of the top trace). All other paths are longer by comparison. Therefore, the current through the via divides inversely proportional to the path length.

The current density on the *top* (trace) layer results in a very interesting pattern (see Figure 9.4). The density is higher at the front edge and lower at the back edge, which might be expected. But the current density totals to the equivalent of almost 4.4 amps (instead of the 3.92 amps measured in the via cores themselves). That is because not all the current at the front edge of the via actually enters the via at the front edge! Part of it continues past the front edge and then enters the via, for example, at the sides. If we think of current density as representing tiny packets of current, some of the packets get counted two (or perhaps even more) times. (This effect will be more pronounced when we look at the case of multiple vias.[2])

When looking at Figure 9.2 it is tempting to compare the current patterns on the trace to water flowing along a trace and then into a hole in the trace. Now we are well aware of the negative attitudes many electrical engineers have toward hydraulic analogs of current flow. Nevertheless, the resemblance is pretty striking.

When thinking of current density as little packets of current that may get counted multiple times on the trace layer, this effect extends a small distance down into the via itself. We ran another simulation that, in effect, separated Core 1 into three individual cores with thicknesses 15, 15, and 200 μm, respectively (approximating the 230 μm of Core 1). Recall that the trace thickness on the top layer is 35 μm. These results are summarized in Appendix F.3. Note how there is still a tendency for the total current (as

Position	density
1	270.7
2	210.9
3	118.0
4	211.7
total	811.3
average	202.8
x area	4.397 A

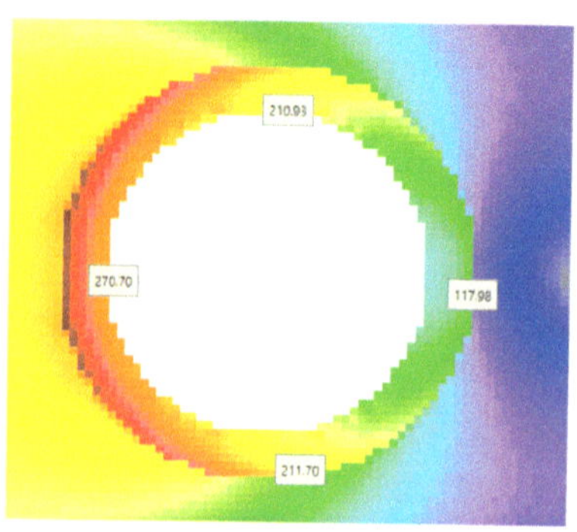

Figure 9.4 Current densities on the top layer.

calculated by multiplying average current density by area) to exceed 3.92 amps even in the second 15-μm layer. After that, the densities are about the same as they were in the previous simulation.

9.4 Multiple Vias

We then simulated a case with four vias arranged in the pattern shown in Figure 9.5. In this simulation the trace width is 0.9 mm (35 mil) and still 34 μm thick. The current in this simulation was still 4.0 amps.

The current density patterns of the first two layers (trace layer and Core 1) are shown in Figure 9.6. The full set of current density patterns by layer is provided in Appendix F.4. The patterns are symmetrical around the horizontal center line. Therefore, we can take the current density readings for either the upper via (in the figure) or lower via; the other via will be the same.

We would expect the current to divide evenly among the four vias. Looking at the current densities in the various core layers confirms that this is true. Each via is carrying (approximately) 0.98 amps. The total calculated current is 3.94 amps (compared to 4.0 amps modeled and the 3.92 calculated with the single via case). (Again, the total does not equal 1.0 amp each because we are modeling a circular via wall with square thermal pixels. Again, see Section F.1.1, "Caution") The current pattern through each via wall looks very similar to that for the single via case. The density is highest at the front of the via and lowest at the back of the via on the outside layers, and uniform around the via wall at the midpoint. That is because the shortest path length through the via is as before; enter the via at the front, travel around the barrel as we go down the via, and out the front with respect to the bottom trace layer. The total current path through each via is identical for each of the four vias.

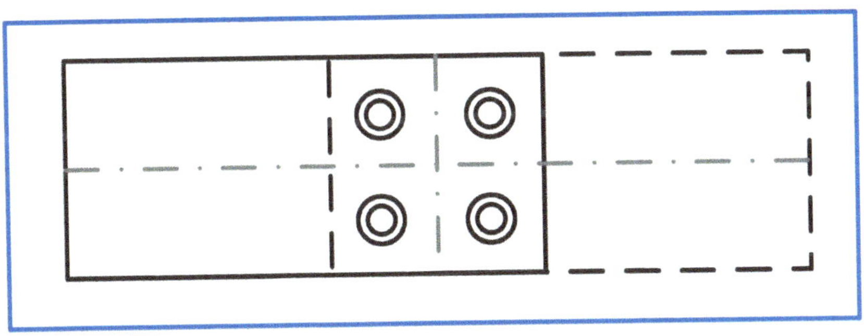

Figure 9.5 Simulation of multiple vias (not to scale).

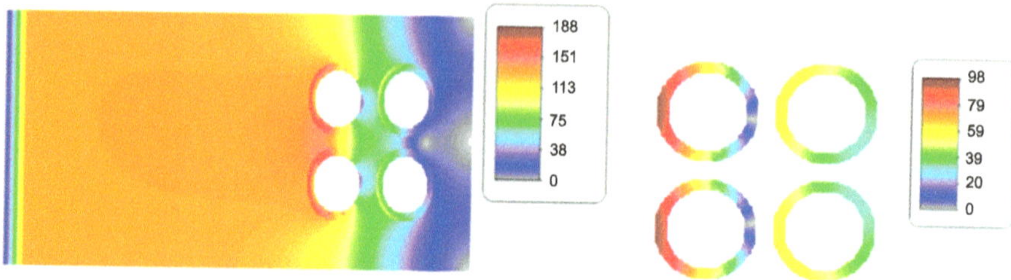

Figure 9.6 Current density for trace layer (left) and Core 1 layer (right).

But the current densities on the *top* (trace) layer are more interesting than before (see Table 9.1). The total current around the front via (found by multiplying the sum of the densities by the via conducting area) is almost 2.3 amps (instead of the 1.0 amp flowing through the via). As before, this is because some of the current seen at that point is flowing into the via, but some is flowing past the front via to the rear via. Using the analogy of little current packets, the incidence of double counting current packets is much greater in this simulation. The data in Section F.4 suggests that this tendency is still slightly apparent in the front vias in the first core layer where the total calculates to 1.1 amps and then 0.98 amp in succeeding layers.

9.5 Multiple Vias and Turn

In our final simulation we used the same trace dimensions as the previous simulation, but we turned the bottom trace 90 degrees (see Figure 9.7).

Table 9.1
Current Density Data on the Top (Trace) Layer for the Four-Via Simulation*

Trace Layer (L)		**Trace Layer (R)**	
Position	**Density**	**Position**	**Density**
1	136.87	1	76.4
2	129.2	2	45.1
3	33.77	3	29.6
4	124	4	66.8
total	423.84	total	217.9
average	105.96	average	54.475
x area	2.299 A	x area	1.182 A

*The front vias are on the left and the rear vias on the right.

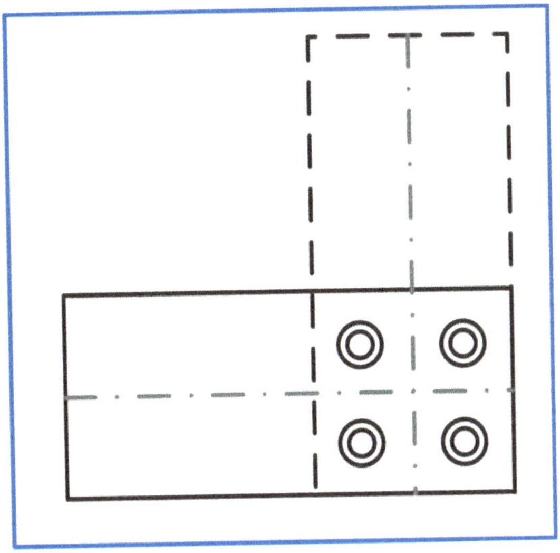

Figure 9.7 Simulation of four vias with traces turned 90 degrees.

The current distribution in the trace layer and in the first core layer are shown in Figure 9.8. The full pattern of current densities through all the layers is provided in Section F.5 in Appendix F.

By now we can almost predict what will happen. The shortest current path will be through the upper left (in Figure 9.8) via, and the longest path through the lower right via. So we would expect the current to divide unequally through this configuration. In fact it does, with the upper left via carrying 1.14 amps and the lower right via carrying 0.72 amp. The other two vias each carry 0.96 amp. Beyond that, the current densities in the via, as we progress down through the layers, are pretty much as in the prior simulation.

9.6 Conclusions

We offer the following thoughts about the foregoing material:

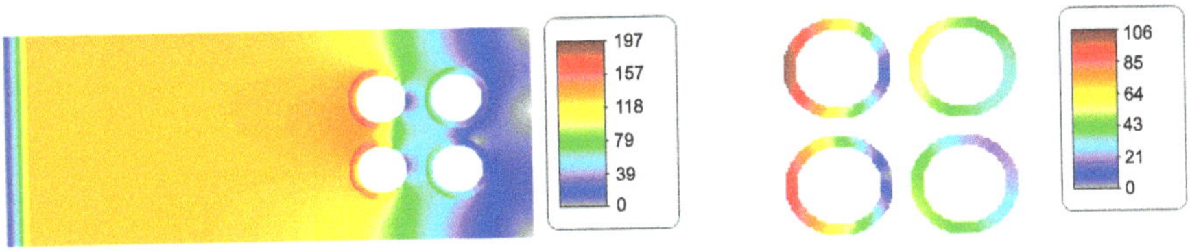

Figure 9.8 Current density for trace layer (left) and Core 1 layer (right).

1. The current density gradients do not contribute in any way to trace and via temperature. The thermal conductivity of the copper in the traces distributes the heat away from any potential point heat generation.

2. TRM results in a steady-state (DC) analysis. We are not aware of any high-speed signal integrity inferences we can make from these observations.

3. Frankly, while these results are interesting, we can't think of any particularly practical import to them! The current density variations are small enough that the localized voltage variations are minimal, and there are no other areas we can think of where these variations might pose a practical problem. We leave it to others to explore this thought.

End Notes

1. Nonuniform current densities are common, for example, when dealing with higher frequencies—the skin effect is one familiar situation.

2. This phenomenon was readily apparent and repeatable in via current density simulations where the via turned into the *z*-axis. It was not apparent in any trace current density simulations when a trace Y'd into another branch on the same plane.

CHAPTER

10

Contents

Thinking Outside the Box

10.1 Bottom Line

- High-current traces do not have to follow any rigid rules. They don't need to be constant width. They don't need to be as wide as we would otherwise think everywhere along their length. We can use short stubs to pass between pads. We can take advantage of underlying traces and planes along portions of the traces. In other words, we can start thinking creatively about how to route them. There are an infinite number of design options we can consider.

10.2 Start Thinking Outside Our Boxes

In Chapter 7 we showed how big an influence on trace temperature an underlying plane can have. And in Chapter 8 we showed how we can carry a substantial amount of current through a via without the via heating much above its parent trace. That is, how a trace can act as a significant heat sink for a via. We now offer one more observation about board design: *a power trace carrying a large current does not need to have a uniform width along its entire length.* Board designers sometimes get so comfortable with trace widths that they hardly think much about them once the required width is defined. Traces tend to be the smallest width the fabricator can comfortably handle in order to conserve real estate or some defined width to meet some controlled impedance target. We almost never think of varying the width of a trace at seemingly random places along its length. What if we thought outside of these boxes?

10.3 Test Board

In the examples to follow, we will assume a standard test board that is 220 mm long by 20 mm wide (8.7 × 0.8 in). Its thickness will be 1.55 mm (approx. 61 mil), including 38-μm (1.0-oz) trace layers top and bottom. A stackup is provided in Figure 10.1. The standard (reference) trace on the board's top layer will be approximately 150 mm (6.0 in) long by 1.0 mm (40 mil) wide. The board dielectric material will be polyimide. We will be applying 4.0 amps of current through the trace. We will simulate our examples and calculate temperatures using TRM. The temperature of the simple trace, with no adjacent traces or planes and with no supplemental cooling, rose from an ambient of 20°C to 53.2°C, or a change of 33.2°C.

10.4 Copper Under the Trace

Adding copper under the trace will help conduct heat away from the trace, lowering the trace temperature. We previously showed in Section 7.2.6 that

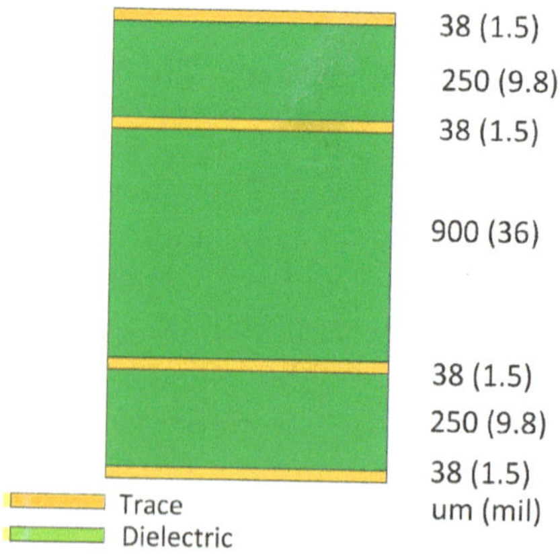

Figure 10.1 Stackup of test board (not to scale).

planes can do this quite well. But if there is not enough room for a plane, perhaps we can get some benefit from simply placing a copper trace underneath our target trace. We looked at five configurations summarized in Table 10.1 and in Figure 10.2. These included

(A) The trace with no supplemental cooling;

(B) Placing a trace the same width on the trace layer under the trace;

(C) Placing a trace 3× wider than the trace (120 mil) on the trace layer under the trace;

(D) Placing a full copper plane on the bottom trace layer;

(E) Placing a full plane on the trace layer under the trace.

10.4.1 Discussion

There are three conclusions that are pretty intuitive. The more copper we can place under a trace, the more heat is conducted away from the trace, and the cooler the trace temperature will be. The closer the copper is to the trace, the more heat is conducted away from the trace, and the cooler the trace temperature will be. The more heat that is conducted away from the trace, the smaller the trace can be and still carry the target amount of current. So, *look for ways to add copper underneath the trace.*

It should be noted that a copper area used specifically for cooling carries no current whatsoever. And even if there is some small amount of noise signal coupled into it from the power trace, that coupled current should be harmless (and can be dealt with by tying the copper area to the system reference (ground) with a *single* via). On the other hand, if the trace is carrying high-current AC or pulsed current, a signal integrity engineer might need to be consulted.

From a thermal standpoint, an underlying plane works like any other plane. The underlying plane does not have to be a continuous and related

Table 10.1
Basic Data

Case		Temperature (°C)	Delta T	% Delta T Improvement
A	Simple trace	53.2	33.2	0.0
B	Narrow trace under	52.95	32.95	0.8
C	Wider trace under	49.2	29.2	12.0
D	Full plane, bottom	44.8	24.8	25.3
E	Full plane, under	39.5	19.5	41.3

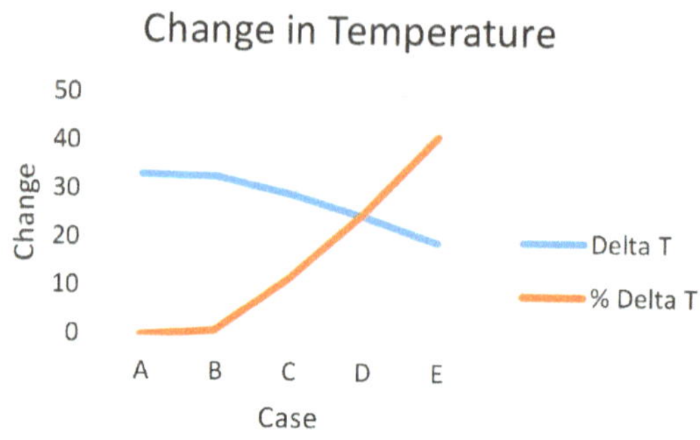

Figure 10.2 Change in temperature data from Table 10.1.

plane in the sense that we need that for signal integrity. However, a systems engineer needs to determine if there is any noise on the high-current-carrying trace that might couple into any underlying unrelated (and sensitive) plane.

However, our simulations showed that adding vias (i.e., thermal vias) connecting traces to planes in any of these configurations resulted in trivial (if any) improvement. Traces can cool vias, but not the other way around. Vias do not cool traces. If there is a parallel plane, the thermal coupling of the trace to the plane through the dielectric is a more effective cooling mechanism than is a thermal via.[1]

10.5 Adding Additional Copper to Traces

Again, it is not necessary for a high-current-carrying trace to be constant width. There may be opportunities to change the width as the trace routes across the board. For this example we added nine stubs or pads, equally spaced along the trace. The stubs increased the width of the trace from 1.0 to 5.0 mm at each point. They were spaced 15 mm apart. Figure 10.3 is an illustration from the simulation output showing the configuration.

With just a simple 1.0-mm trace, the maximum trace temperature was 53.2°C. Adding the distributed stubs lowed the peak temperature to 44.4°C.

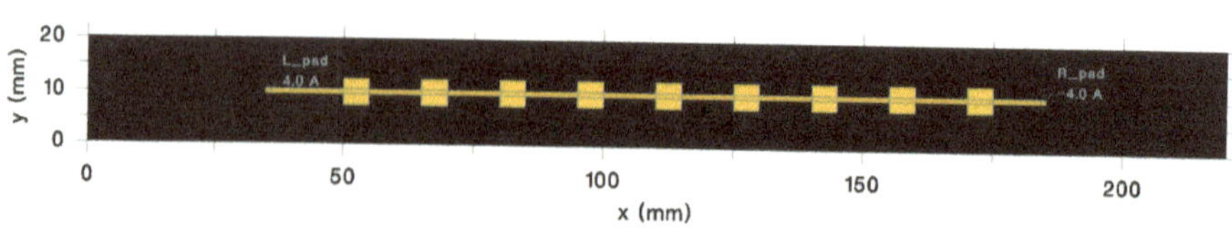

Figure 10.3 Trace with distributed stubs along its length.

The corresponding changes in temperature were 33.2°C and 24.4°C, respectively, an improvement of 26.5%. This means we could, for example, decrease the width of the standard trace without exceeding our maximum allowable trace temperature. The thermal profiles of the two cases are shown in Figure 10.4.

10.5.1 Discussion

If we do not constrain ourselves thinking that high-current-carrying traces should be a constant width, we open up all kinds of opportunities to change the width where we can to help conduct away heat. So if, for example, we calculate that we need a trace that is 40 mils wide to carry the current, we may be able to reduce that width in some areas of the board while increasing it in other areas without increasing our maximum tolerable trace temperature.

The largest part of the current is carried along the line of the trace. There is a little current fringing into the stub, not very much. Figure 10.5 shows the current density along a portion of the trace.

10.6 Dealing with Connecting Links

Short, narrow connecting links along high-current-carrying traces are not as troublesome as we might think. The main trace acts as a heat sink for the link, conducting a significant part of the heat away from the link, helping to lower the temperature of the link. In this example we will look at three link configurations (see Figure 10.6):

(A) A single, short, 3.0 mil long, 11 mil wide connecting link between two sections of our test trace. Such a link may be required if we

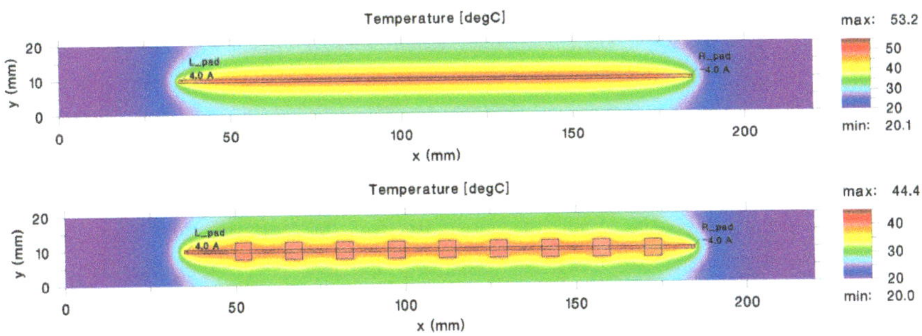

Figure 10.4 Thermal profiles showing the simple trace and a trace with numerous stubs along its length.

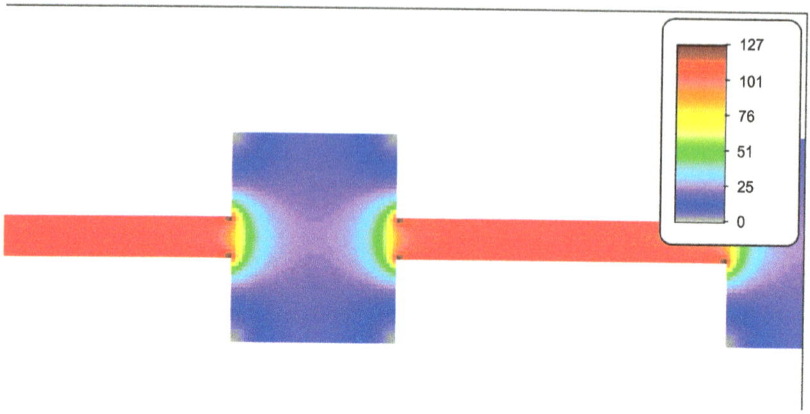

Figure 10.5 Current density along trace (units are A/mm2).

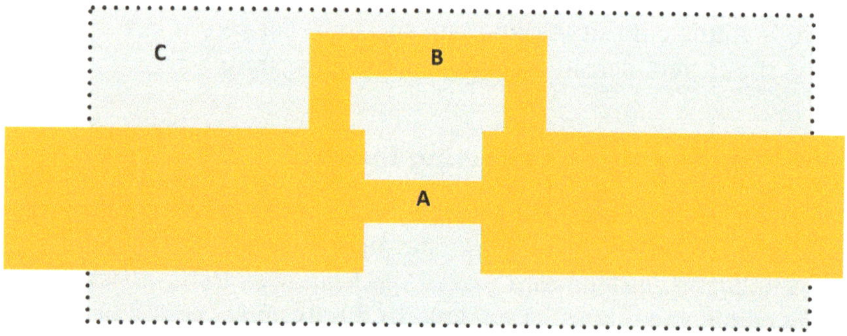

Figure 10.6 Three link configurations, not to scale.

have to bring our primary trace near some components that restrict the area around the trace. This link looks a lot like a fuse!

(B) The addition of a second link using a nearby path.

(C) The addition of a small, supplemental area of copper (partial plane) directly underneath a single link. This copper area is 40 mm long by 6.0 mm wide (1.6 × 0.25 inches).

The results of these simulations are shown in Table 10.2. A single, 11-mil wide trace by itself carrying 4.0 amps would heat to well over 150°C. But because of the heat-sinking influence of the basic trace, the trace temperature doesn't rise nearly that much. But it does increase, perhaps to intolerable levels. If we put a small plane under it, the temperature only increases about 13%. Adding a parallel second link, the benefit is about the

Table 10.2
Example 3 Data

Case	Temperature	Delta T
Basic trace	53.2	33.2
Single link, A	74.6	54.6
Single link + plane C	60.0	40.0
Second link B	59.3	39.3

same as adding an underlying plane. The thermal profiles of the three cases are shown in Figure 10.7.

10.6.1 Discussion

If we are constrained in our thinking that high-current-carrying traces must be a certain width, and we then run into a constricted area on the board, it would seem that we had few, if any options. But if we relax that restriction, several opportunities may present themselves. It certainly may be possible that a second (albeit narrow) adjacent path (or even more) might be available. This would allow spreading out the heat generating area of the trace over a wider area, helping to control the trace temperature.

The plane area under the link area need not necessarily be continuous. In fact, it need not even be a single area. There might be an opportunity to spread copper around the layer between other traces. And any via paths that might exist through the plane are almost inconsequential as far as the heat-spreading ability of the copper is concerned.

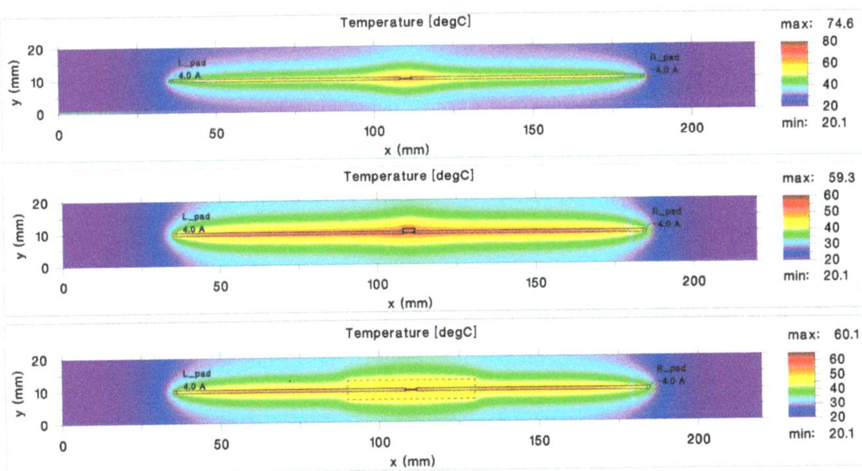

Figure 10.7 Thermal profiles of cases A, B, and C (top to bottom).

10.7 Conclusions

When we start worrying about signal integrity issues in PCB design, we start worrying about *rules*: loop area must be minimized; trace impedance must be constant; pad impedance must be minimized. We are now advocating that when designing high-current-carrying traces we forget any rules (i.e., the trace must be a minimum size). Instead, start thinking outside the box. Think about "How can I lower the trace temperature in ways other than just increasing its cross-sectional area?" As a start, we can add (noncurrent-carrying) copper areas near and under the trace to help conduct heat away. We can design our traces with wider widths where we have room and narrower widths where we don't. The wider areas will help sink some of the heat away from the narrower areas. In this way we will achieve much more flexibility in our layouts, especially in tight areas.

The good news from all this is that there may be some really significant options available to us if we look. The bad news is that it is really difficult to evaluate them! Charts and formulas are not going to help much here. It would appear we can only evaluate concepts like these using thermal simulation models.

End Notes

1. This comment applies to traces, which have a certain surface area in contact with the dielectric. If we have a small component that generates heat in a very small area (think "spot"), then thermal vias might be an effective way to conduct heat away from that hot spot.

CHAPTER

11

Contents

Fusing Currents: Background

11.1 Bottom Line

- W. H. Preece and I. M. Onderdonk were the first people credited with looking at how much current a copper wire can carry before reaching the melting point. They developed Preece's equation and Onderdonk's equation, respectively.
- Onderdonk's equation explicitly includes time as a variable, but excludes any cooling effects. Consequently, it is only valid for a few seconds.

11.2 Background

At the end of a seminar one day, a participant asked the following question:

"If there is a catastrophic failure of my system, a certain 1.0-oz trace will carry 40 amps. I need it to carry that for 1.0 second while the system shuts down in a controlled manner. What size trace do I need?"

The first time this question was asked we referred the person to people in the fusing industry. The second time we went to the people in the fusing industry ourselves and found that they didn't have a clue how to proceed!

It turns out the early work on this topic goes back to W. H. Preece in the 1880s, and I. M. Onderdonk, possibly in the 1920s. Preece's work has been well documented, although until recently copies of that documentation have been difficult to obtain. There is growing speculation, however, that Onderdonk may never have published his source data under his/her own name.

In this chapter we will (a) discuss the background of Preece, his motivating factors, and links to his source data; (b) discuss what can be inferred about Onderdonk, his likely motivating factors, and links to early work based on his equation; and (c) outline the critical assumptions behind these formulas. In the absence of any Onderdonk source data, we independently derive Onderdonk's equation in Appendix G.

11.3 W. H. Preece

In the 1880s, Sir William Henry Preece was a consulting engineer for the British General Post Office, and in 1892 became engineer-in-chief.[1] At that time, the post office was responsible for the telegraph (and later wireless telegraph) system in England. He published three papers[2] in *Proceedings of the Royal Society of London* in the 1880s that formed the basis for his famous equation:

$$I = a * d^{3/2} \tag{11.1}$$

where d is the diameter of the wire in inches, a is a constant (10244 for copper), and I is the fusing current in amps. A little algebra transforms this equation to

$$I = 12277*A^{3/4} \tag{11.2}$$

where A is the cross-sectional area in square inches.

Interestingly, in the 1890s Preece became an ardent supporter, both politically and financially, of Guglielmo Marconi and his work on transoceanic wireless telegraphy.

Preece was motivated by the problem of lightning striking telegraph wires. The early form of lightning arrester used in that period is shown in Figure 11.1, reproduced from Preece's 1883 paper.[2] Most of the energy from the strike was intended to flow through the capacitor, but some was left to travel onto the cable, which led to the telegraphy equipment. Preece

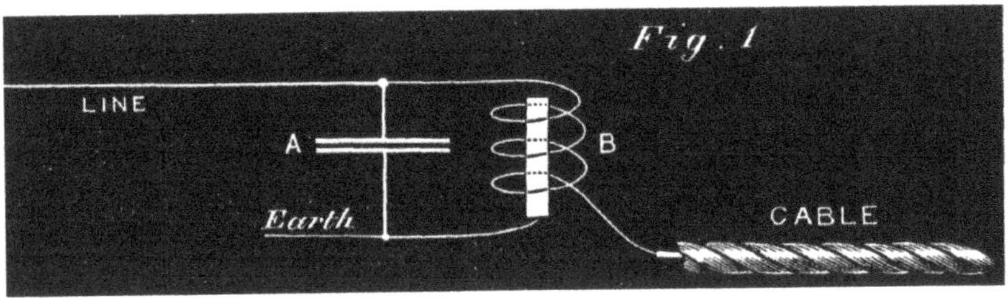

Figure 11.1 Early form of lightning arrester. (Source: 1883 paper).

wanted to determine the best material, and its size, for the lead from the arrester to the cable. The wire had to be able to carry the normal current for telegraph operations, but fuse at currents slightly above that level in order to protect the operators.

Preece did not approach this problem analytically. He was, what we might call today, a lab rat, and did all his experiments in the lab. He would apply current to a test wire until the wire began to glow (but not yet melt). He would call that point the fusing point (more appropriately the melting point). He looked at a large selection of materials and provided the constant (a in (11.1)) for each material. He concluded that "the best metal to use for small diameters was platinum, and for large wires tin (see end note 2, 1887 paper). The final constants he derived were summarized in his 1888 paper,[2] which is where the value 10244 for copper is found.

11.4 I. M. Onderdonk

I. M. Onderdonk is much more obscure. Speculation is growing that Onderdonk never published his work under his own name, and we have uncovered no citations for any work by Onderdonk himself. His equation is offered almost as a given, in much the same way we reference Ohm's law today. His equation appears in several papers and sources,[3–7] usually in the form

$$33\left(\frac{I}{A}\right)^2 S = \log_{10}\left(\frac{t}{274}+1\right) \tag{11.3}$$

for $T = 40°\text{C}$, but occasionally in the more general form:

$$33\left(\frac{I}{A}\right)^2 S = \log_{10}\left(\frac{t}{234+Ta}+1\right) \tag{11.4}$$

where

I = the current in amps

A = the cross-sectional area in circular mils[8]

S = the time in seconds the current is applied

t = the rise in temperature from the ambient or initial state[9]

Ta = the reference temperature in °C

Some publications erroneously cite Onderdonk's name in a citation (see end notes 4, 5).

End note 4 is an anonymous (not by Onderdonk, as some citations suggest) single-page nomograph inserted in *Electrical World* magazine. A link is provided to it in end note 4. There are notes on the face of the nomograph that say (somewhat cryptically, and among other things) "...based on an equation by I. M. Onderdonk," and "Chart from engineering department of a midwestern utility."

References to Onderdonk's equation frequently include a nomograph. Before the age of computers, a nomograph was a common way to represent a complex problem for graphical solution. Figure 11.2 represents the type of nomograph that commonly accompanies Onderdonk's equation. The horizontal axis is current and the vertical axis represents time. The red diagonal lines represent various types and sizes of electrical cable. Each point along the line representing each particular cable type and size and represents a solution (current and time) to Onderdonk's equation.

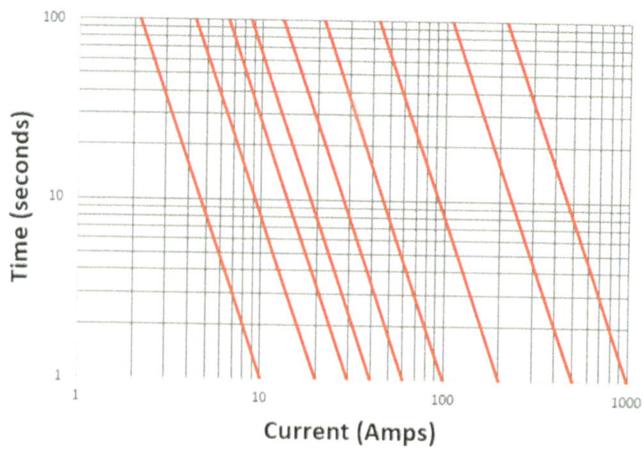

Figure 11.2 Typical type of nomograph for solving Onderdonk's equation.

End note 5 is a respectable article written by E. R. Stauffacher (but *not* coauthored by Onderdonk). It contains a different nomograph. Stauffacher comments that the "Values of short-time current required to melt copper conductors, as given by the chart shown in Fig. 1, were calculated from the formula..." and then he offers Onderdonk's equation. A footnote to the equation says "This formula was developed by I. M. Onderdonk."

It is tempting to speculate that, since Stauffacher is the earliest reference we have found, Onderdonk may have been a contemporary of Stauffacher. Stauffacher, himself, was employed as Superintendent of Protection at Southern California Edison when he wrote his article (in 1928). Perhaps Onderdonk was an engineer at SCE at the same time.

Stauffacher's article offers a clue as to what the motivation was for Onderdonk's equation. In approximately that same time frame, SCE was building a high-power transmission line from Southern California to the construction site for Boulder Dam (now Hoover Dam) in Nevada. The concern was that under certain conditions there may be a short circuit that occurs between a high-voltage transmission line and the copper wires attached to the insulators or poles supporting the line. These copper wires needed to carry that short-circuit current for the short period of time "required to clear the faulty line from the system."[5] Thus, the desirability of having a variable for time (S) in the equation.

11.4.1 Cautions

It is important to understand what constitutes the "melting" of a copper wire. When we apply heat to a wire (or any solid for that matter), there are two stages we can consider. The first is that the wire heats up from the initial temperature to the melting temperature. Let's call this time $t1$. Then additional heat (and time, say time $t2$) is required to change the wire from a solid to a liquid (i.e., to actually melt the wire). This second amount of heat is referred to as the heat of fusion.[10] Time $t1$ is required to heat the wire *to* the melting temperature, and time $t1 + t2$ is required to actually melt the wire from the initial temperature.

Even if the trace is beginning to melt, current may flow through the liquid copper. However, liquid copper has a lower electrical and thermal conductivity than does solid copper (refer back to Table 3.1 and Figure 3.3). Therefore, once the liquid starts to form, there may be an explosive runaway condition that follows. The circuit will break (i.e., current will stop flowing) only if the liquid path separates. This may happen as a result of gravity, as a result of constricting surface tension, or as a result of explosive splattering. Therefore, if we consider time $t2$ to be the time the circuit path opens, that can depend on many subtle variables.

Preece clearly refers only to time *t*1. He raised the current *to* the point where the wire glowed. Onderdonk's equation also only applies to time *t*1. This is inferred from most (but not all) of the references and is confirmed by the derivation in Appendix G.

Unfortunately, the semantics found in some of the articles seems to confuse this point. For example, the nomograph in *Electrical World* is titled "Short-Time Current Required to Melt Copper Conductors" (emphasis on "melt"). Stauffacher refers to the "short-time current required to melt copper conductors." But since both are based on Onderdonk's equation, they actually only refer to the time to bring the temperature *to* the melting point.

Preece's equation is based on slowly bringing the current up to the melting point. But he did not bring the variable of time into the equation. There would have been some cooling effects that occurred while he was heating the various materials, but this was of little consequence in what he was trying to discover.

Onderdonk, on the other hand, specifically brought time into the relationship, and therefore he had to make some assumption about the cooling effects that might have coexisted with the heating effects. His assumption was that there were no cooling effects (from conduction, convection, or radiation). Thus, the heating was rapid (as might happen in a short-circuit situation). That is why the references say that Onderdonk's equation is only valid for short periods of time, typically less than 10 seconds for wires suspended in the air. People can quibble about this time. For example, we show in Chapter 12 that cooling effects in PCB materials can significantly impact Onderdonk's predicted fusing time in as little as 1.0 second. In any event, Onderdonk's equation becomes less and less reliable as time increases.

End Notes

1. http://en.wikipedia.org/wiki/William_Henry_Preece.
2. Preece W. H., "On the Heating Effects of Electric Currents," Proc. Royal Society, Vol. 36, 1883, pp. 464–471; No. II, Vol. 43, 1887, pp. 280–295; No. III, Vol. 44, 1888, pp. 109–111, http://rspl.royalsocietypublishing.org/content/36/228-231/464.full.pdf+html; http://rspl.royalsocietypublishing.org/content/43/258-265/280.full.pdf+html; http://rspl.royalsocietypublishing.org/content/44/266-272/109.full.pdf+html, or at www.ultracad.com/articles/reprints/preece.zip.
3. Hink, D. G., and H. W. Beaty, *Standard Handbook for Electrical Engineers,* 14th Edition, 1999, McGraw-Hill, p. 4-74.
4. Anon., "Short-Time Current Required to Melt Copper Conductors," *Electrical World,* Vol. 121, No. 26, June 24, 1944, pp. 98 (download a copy from www.ultracad.com/articles/reprints/electrical_world.pdf).

5. Stauffacher, E. R., "Short-Time Current Carrying Capacity of Copper Wire," *General Electric Review,* Vol 31, No. 6, June 1928 (download a copy from www.ultracad.com/articles/reprints/stauffacher.pdf).

6. Babrauskas V., and I. Wichman, "Fusing of Wires by Electrical Current," *Fire and Materials 2011 Conference Proceedings,* Interscience Communications (download a copy from https://www.academia.edu/9357019/Fusing_of_Wires_by_Electrical_Current or from www.ultracad.com/articles/reprints/babrauskas.pdf).

7. Codreanu, N., R. Bunea, and P. Svasta, "New Methods of Testing PCB Traces Capacity and Fusing," Note: Their copy of Onderdonk's formula is incorrect and contains a misprint! (Download a copy from www.ultracad.com/articles/reprints/codreanu.pdf).

8. A circular mil is the area of a circle with a diameter of 1 mil. The formula is $A = d^2$. The conversion from circular mils to mil^2 is $\pi/4$. Normal conversions are

   1 mil^2 = 1.273 circular mils

   1 circular mil = .7854 mil^2

   1 m^2 = 19.736*10^8 circular mil

   1 circular mil = 5.067*10^{-10} m^2 = 5.067*10^{-4} mm^2

9. From a thermodynamic engineering standpoint, there is a significant difference between the ambient temperature and the initial condition. For example, there could be some heating of a wire above the ambient temperature in a normal operation before a change of current is applied. In this text we are assuming that the initial condition and the ambient temperature are the same.

10. http://en.wikipedia.org/wiki/Melting_point.

CHAPTER

12

Contents

Fusing Currents: Analyses

12.1 Bottom Line

- Our fusing simulation of Onderdonk's equation fits his equation so closely you can't tell the curves apart.
- Onderdonk's equation assumes there is no cooling. On a circuit board this assumption is absolutely wrong. Therefore, simulations and experiments on real-world boards follow Onderdonk's equation for only approximately 1 second.
- Nevertheless, we can adjust Onderdonk's equation results to get reasonable approximations for real-world situations for fusing times within about the first 5 seconds.

12.2 Background

In Chapter 11, Section 11.2, we recalled a student's question and then covered the historical background investigations into this topic. Of relevance to us is Onderdonk's equation :

$$33\left(\frac{I}{A}\right)^2 S = \log_{10}\left(\frac{\Delta T}{234 + Ta} + 1\right) \tag{12.1}$$

A more useful form of Onderdonk's equation (at least for us) is (12.2), in which we rearrange the terms to solve for t (time, which is S in (12.1)):

$$t = \left(\frac{1}{33.5}\right)\left[\log_{10}\left(\frac{\Delta T}{234 + T_{ref}} + 1\right) * \left(\frac{A}{I}\right)^2\right] \tag{12.2}$$

The various variables are

I = the current in amps

A = the cross-sectional area in circular mils[1]

$S = t$ = the time in seconds the current is applied

ΔT = the rise in temperature from the ambient or initial state[2]

Ta = the reference temperature in °C

In this chapter we are going to look at the question of trace fusing currents using Onderdonk's equation and also some thermal simulation models of traces.

12.3 Fusing Time and Temperature

It is important to understand what we mean by fusing temperature. We covered this in detail in Chapter 11 and we refer you back to Section 11.4.1. Fusing temperature in our case is the melting temperature of copper, 1083°C. Fusing time is the time to reach the fusing temperature (but not the additional time necessary to actually melt the trace.

12.4 Assumptions and Cautions

We are talking about the fusing temperature of PCB traces. We are not talking about fuses per se. So, for example, we are not talking about a commercial fuse of the type shown in Figure 12.1(a) nor are we talking about forming a fuse link along a PCB trace of the type shown in Figure 12.1(b)

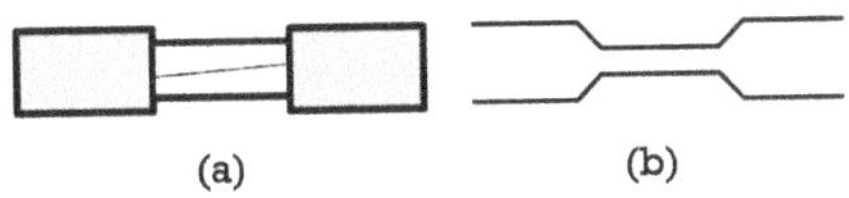

Figure 12.1 Types of fuses we are not considering.

(although the principles we explore here probably apply equally well to Figure 12.1(b)). Nor are we talking about an insulated wire. Instead, we are talking about standard PCB traces of uniform width and uniform thickness over their relevant length.

12.5 Simulation Models

In Chapter 6 we developed a thermal simulation model of a 6.0-inch trace and used that in various sensitivity scenarios. We will continue to use a similar simulation model in this chapter. This model is referred to in our graphs as "TRM Trace." The various forms of this model range in thickness from 1.0 to 2.0 oz, and in width from 20 to 200 mil. The trace model is placed on a 63-mil thick FR4 substrate.

We also developed a new model for this chapter we call TRM Fuse" It is a solid strip of copper, 200 mil long and also varies in width from 20 to 200 mil. It varies in thickness from 1.0 to 2.0 oz. Since TRM is not designed to model a single layer structure, we modeled the fuse on two copper layers as shown in Figure 12.2.

Onderdonk assumed no cooling processes in the derivation of his equation. To model that kind of fuse we had to eliminate any heat paths associated with conduction, convection, and/or radiation. In our model, both conductor layers are copper and there is no dielectric layer. The heat exchange coefficient (see Chapter 6) was set to zero (but setting it to 10 had negligible effect on the simulation model results because the time frames are so short).

There is a subtle detail involved in running a simulation of this type. In order to get a precise measure of fusing time, the model needs to run for a

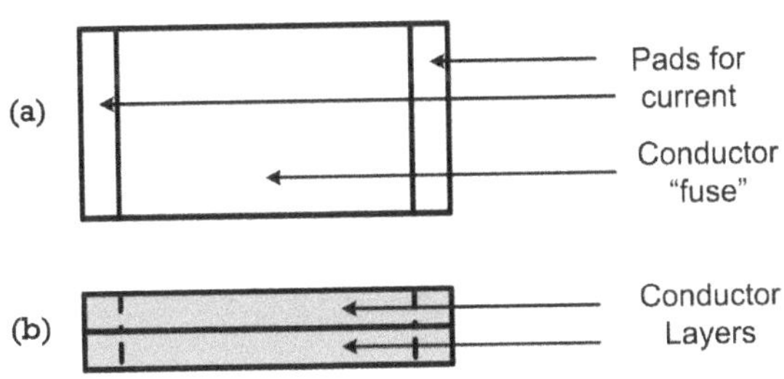

Figure 12.2 Simulation model of TRM Fuse.

period of time, calculate the temperature of the copper, adjust the resistivity of the copper to the new temperature, and then run the simulation for the next period of time. This continues until the melting point of copper (1083°C) is reached. Each period of time is called a step. The step size impacts the results; the most precise results are reached with infinitely small steps (i.e., in the limit where step approaches zero). A step size that small results in very long calculation times. After considerable experimentation we found that setting the step to 0.1 second was a reasonable compromise.

12.5.1 Simulation Results, TRM Fuse

We have expressed Onderdonk's equation in the form for solving for time in (12.2). In that equation, *A* is still in units of circular mils. If we assume a reference temperature of 20°C, a fusing temperature of 1083°C, and convert this to square mils, the result is (12.3):

$$t = .0346*(A/I)^2 \tag{12.3}$$

We can plot this equation as a curve and it represents the theoretical fusing time for a conductor (as based on Onderdonk's equation). Then if we run a TRM Fuse simulation at various currents and plot the results of the simulation, we can compare the simulation model results with the theory. Figure 12.3 is the result for a 1.0-oz, 100-mil wide trace.

The simulated results lie practically on top of the theoretical results! It is very difficult to tell the two curves apart. The fit could not be better. This

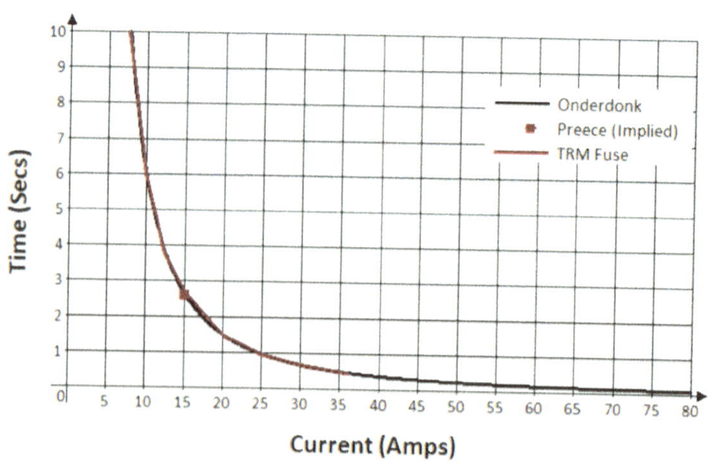

Figure 12.3 Simulation results, 100-mil, 1.0-oz trace, against Onderdonk's equation.

gives us a reasonably comfortable feeling that our approach here is viable. The red dot on the curve is the implied point[3] from Preece's equation (see Chapter 11) for the same level of current and cross-sectional area.

Six different simulations were run, 1.0-oz (1.35-mil) and 2.0-oz (2.7-mil) thicknesses with widths of 20, 100, and 200 mil. The results for all other configurations are identical to this one except for scale. The results for all configurations are provided in Appendix H. This gives us a high degree of confidence in TRM's ability to simulate a fuse in this kind of simulation.

12.5.2 Simulation Results, TRM Trace

We constructed a trace model for simulation that was a very simple PCB trace, 155 mm (approximately 6.0 inches) long. It varied in width and thickness depending on what we were testing. There were 3.0 mm (118 mils) between each side of the trace and the edge of the board, and the ends of the trace were 5.0 mm (200 mils) in from each edge. A schematic of one of the traces is shown in Figure 12.4, which is a figure from one of the TRM Trace simulations.

The same six configurations as described above were run on this trace model. The 1.0-oz, 100-mil wide simulation result is shown in Figure 12.5. (The results for all six configurations are shown in Appendix H.)

There are three types of cooling that occur when these traces heat—conduction into the board material, convection into the air, and radiation away from the trace. The heat exchange coefficient was set in the simulations to 10 to represent a normal situation. Subsequent testing showed that the value of the Heat Exchange Coefficient did not change the results significantly, showing that the primary cooling effect in the very short time frames looked at here is confined to conduction through the board material. The significant shift of the trace model simulation to the right of Onderdonk's equation is the result of conduction of heat into the board material, thereby slowing the increase in temperature.

For very high currents, and therefore very short fusing times, the curves are fairly close to each other (at the far right-hand side of the curves). But cooling effects kick in fairly quickly, separating the curves. In a time frame as short as 1.0 seconds the trace model can handle significantly more current than can the fuse model.

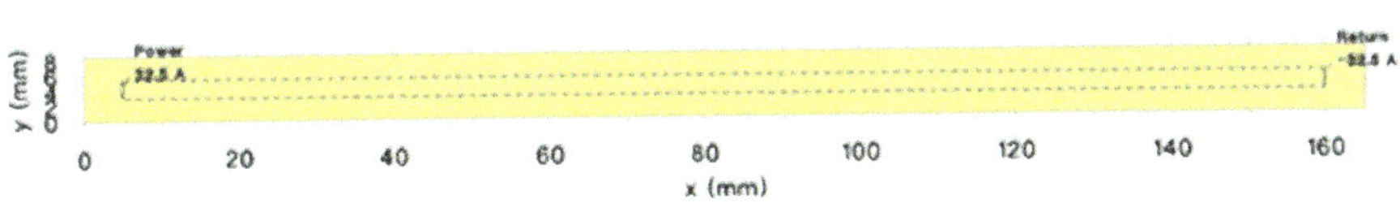

Figure 12.4 Schematic of our TRM Trace model for simulation.

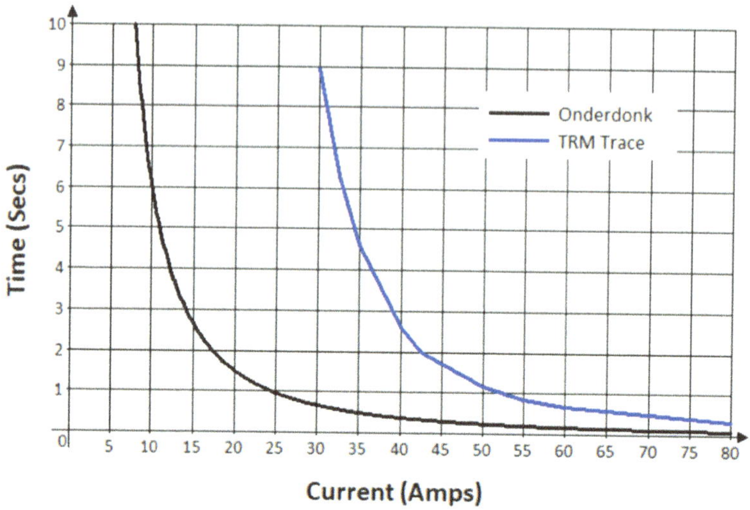

Figure 12.5 Fusing time vs. current for a 1.0-oz, 100-mil wide trace simulation compared to Onderdonk's equation.

The dielectric material used in this simulation was FR4. Polyimide might provide slightly better heat conductivity, better cooling, and therefore slightly longer fusing times. The size of the board had no effect on the results. Placing a plane on the opposite side of the board did not affect the results. Placing a plane 12 mils under the trace layer did have an effect on the results, increasing the TRM Trace fusing time by 30% to 100 % depending on the current level (lower current levels, and therefore longer fusing times, resulted in higher percentage changes).

We also ran a simulation on the trace model we used for simulating the IPC test board (see Chapter 6). The results were *identical* to those from our trace model (within less that 0.1 second at every current level). This gives further confidence in our feeling that the fusing time is a function of trace size and board material only, and nothing else (except a closely spaced underlying plane). This also gives further confidence in TRM's ability to model fusing currents and times.

12.5.3 Short-time Effects

Let's look back at (12.3) and rearrange terms:

$$t(I/A)^2 = .0346 \tag{12.4}$$

This says that the product on the left is a constant.[4] Figure 12.6 plots the result of $t^*(I/A)^2$ for the various fuse model simulations. The results are the same as the Onderdonk result within reasonable measurement accuracy for times greater than 1.0 seconds. There is a slight anomaly for the larger 2-oz traces at the shortest times caused by measurement uncertainties in that region. The current levels are very high and the sensitivity to current level very low in that region for those sizes.

When we plot the same term for the TRM Trace model simulations, we get the interesting result shown in Figure 12.7. They are all upward sloping to the right, a clear indication of the cooling effects that are taking place with time.

But when we look closely at these same curves at very short times, they all seem to originate at a similar value (Figure 12.8). This is not too far different from our fuse simulation. Therefore, we conclude that at the first instant of time, all the trace configurations start out the same. Furthermore, except for the smallest trace (1.0 oz, 20 mil), they are all at or under 0.25 during the first 3 seconds.

It is instructive to compare the fusing current for the traces at various times to the fusing current calculated from Onderdonk's equation. We calculate the ratio of these two values and provide them in Table 12.1. For example, for a 2.0-oz, 20-mil wide trace, the fusing current to cause that trace to fuse in 1.0 seconds is 3.87 times greater than that calculated from Onderdonk's equation. Again, that difference is caused by the cooling effects that are absent in the Onderdonk model.

A very interesting result here is that, relatively speaking, narrower traces take longer to reach the fusing temperature than do wider traces. That

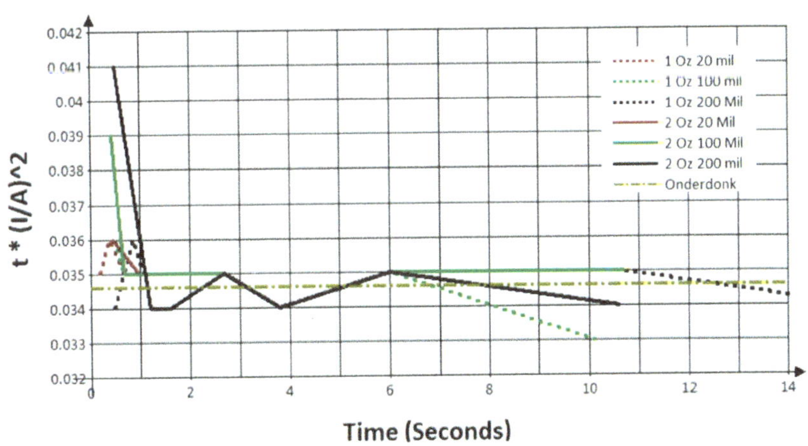

Figure 12.6 Plot of $t^*(I/A)^2$ for the various configuration simulations of the fuse model.

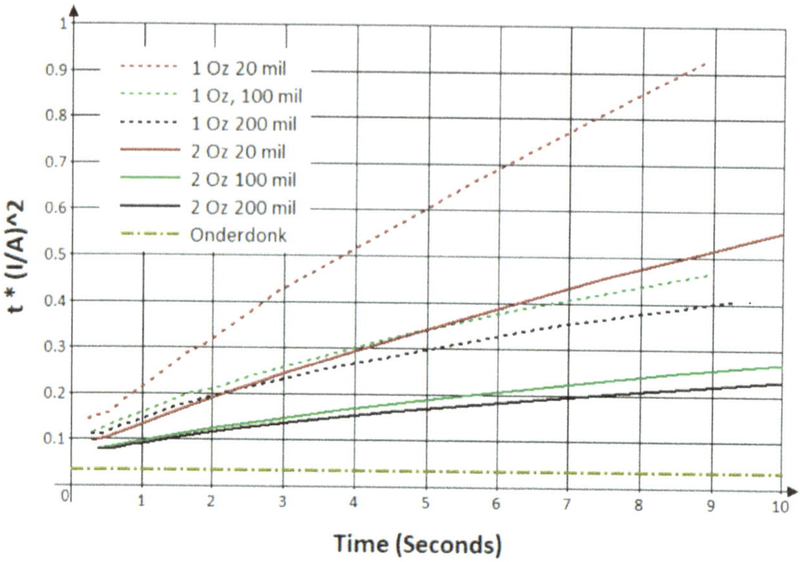

Figure 12.7 Plot of $t^*(I/A)^2$ for the various configuration simulations of the trace model.

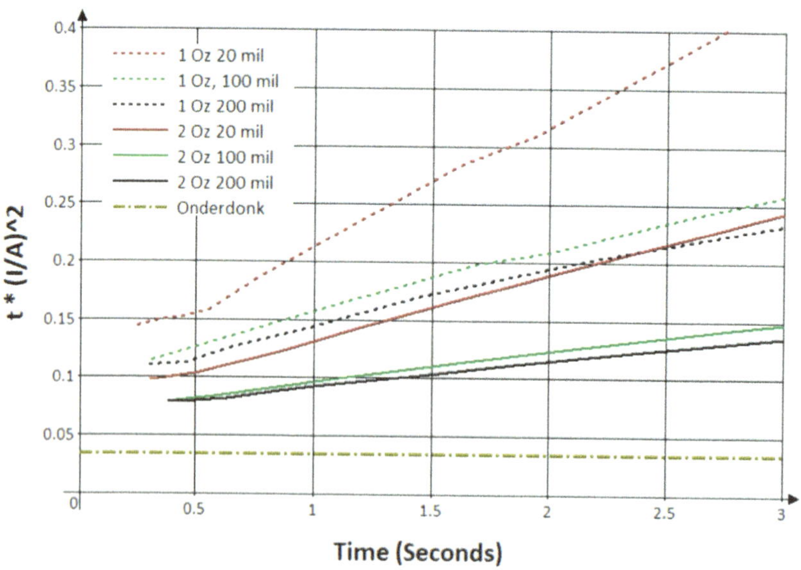

Figure 12.8 Plot of $t^*(I/A)^2$ for the various trace configuration models for times under 3 seconds.

seems, on the surface, to be counterintuitive. But the narrower traces do cool better. Figure 12.9 gives us a clue as to why. The top thermal image is

Table 12.1
Ratio of TRM Trace Model Current to Onderdonk Current

Trace Configuration		Area	Time (Seconds)					
Oz	Width	Mil²	0.5	1.0	2.0	3.0	4.0	5.0
2	20	52	3.46	3.87	4.65	5.30	5.84	6.30
1	20	26	2.24	2.45	3.03	3.53	3.83	4.20
1	100	130	2.02	2.14	2.45	2.69	3.02	3.15
1	200	260	1.88	2.04	2.37	2.62	2.77	2.97
2	200	520	1.57	1.67	1.90	2.05	2.20	2.37
2	100	260	1.51	1.63	1.82	1.98	2.10	2.20

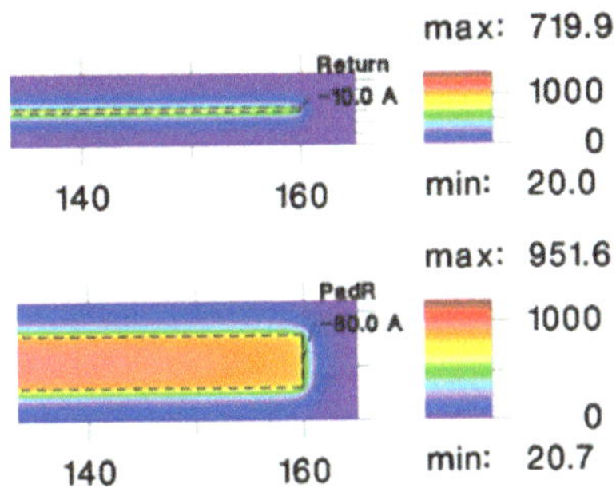

Figure 12.9 Comparing the thermal plume under two traces of differing width.

from underneath the 1.0-oz, 20-mil trace when it reaches the fusing temperature. The lower thermal image is the same thing for a 2.0-oz, 200-mil trace. The cooling plume is the dark blue area around which the trace is shedding heat. Comparing the area of this plume to the trace width illustrates that the narrower trace has, relatively speaking, a much wider plume *compared to its width* than does the wider trace. This is due in part to the time constant associated with how *fast* the heat can flow away from under the trace. (Recall from Section 7.2.2 that it can take 4 to 5 minutes for temperatures to stabilize.) The temperature is therefore cooler under the narrower trace than under the wider trace (720°C compared to 952°C).

Referring back to the question first posed in this chapter (how large a trace is needed to carry 40 amps for 1.0 seconds), we can approximate an answer to that question from Figure 12.8 and Table 12.1.

We assume the trace will be at least 100 mils wide

Calculate $t(I/A)^2 < .15$, or $(1)*(40/A)^2 < 0.15$ or $(40/A)^2 < 0.15$

This leads to: $A > (1600/0.15)^{.5} = 103$ square mils

A 1.0-oz, 76-mil wide trace has an approximate cross-sectional area of 103 sq. mils.

We can check the answer by calculating the fusing current for this result using Onderdonk's equation and then looking at Table 12.1.

Onderdonk: $t = .0346*(A/I)^2$ (from (12.3))

This leads to $I^2 = .0346(A^2)/t = .0346(103)^2$

Leads to $I = 19$ amps

From Table 12.1, a 130 sq mil area trace model will have a 2.14 times greater current carrying capacity, so the current carrying capacity of the trace would be about 2.14*19 = 40.6 *A*, roughly what we wanted.

12.5.4 Final Conclusions

Looking at the question of fusing currents on PCB traces can be very problematic. Onderdonk's equation is a reasonable place to start, and thermal simulation models suggest that the actual fusing times can be 1.5 to 6.0 times longer than Onderdonk suggests (and more considering the restrictive assumptions of the models). But there are many subtle variables involved. It is not really practical to solve this type of problem with a formula or graph. The use of a simulation model is almost required, and even more so if there are adjacent traces or planes that can contribute to trace cooling.

NOTE: If a trace fuses, it is never acceptable to repair the trace and put the board back into service. If a trace fuses, that is considered a destructive failure and the board must be removed from service. The heat from the fusing temperature may have caused any number of hidden potential problems in the material and components around that area. It may be useful to use a trace as a fuse, but doing so is a one-time event. If a replaceable fuse is required, a fusing component must be used instead.

12.6 Experimental Results:

In Section 12.5 we used TRM to simulate the fusing characteristics of a fuse and of a typical PCB trace. The fuse model was pretty straightforward, although pretty theoretical. But when it comes to simulating an actual trace, there are some significant problems we need to be aware of. Almost all of these potential problems are caused by the fact that we are covering a *very* wide temperature range, from around 20°C to 1,083°C. Then, it is one thing to deal with these in simulations. It can be even more difficult to deal with in actual experimental trials.

12.6.1 Heating Uncertainties

Recall that traces heat by I^2R power dissipation in the trace. The change in temperature is given by the same formula we used in Chapter 4 ((4.4)), repeated here as (12.5):

$$\Delta T = \frac{1}{\alpha_0}\left[\frac{R_t}{R_{t_0}} - 1\right] \tag{12.5}$$

where

ΔT = change in temperature from ambient (ref) (°C)

α = thermal coefficient of resistivity of copper (1/°C)

Rt = resistance at the temperature under test (ohms)

R_{to} = resistance at the reference temperature (ohms)

Recall from (3.1) that resistance is given by (12.6):

$$R = (\rho / A) * L \tag{12.6}$$

where

ρ = resistivity

A = area = width * thickness

L = length of the trace

We have already noted that thickness can vary significantly along a trace and from trace to trace. We have already noted that this type of analysis leads to an average temperature for a trace, but there may be hot spots along the trace. And we have already noted that α varies with temperature and we may not know exactly what it is at any point along a trace. Therefore, it should be no surprise that applying our models to a specific trace involves some faith that we actually know what we are dealing with.

12.6.2 Cooling Uncertainties

A PCB trace typically cools primarily by conduction through the dielectric. It also cools through convection and radiation. Computer models are reasonably good at handling the conductive cooling as long as the thermal properties of the material are known and predictable (i.e., as long as the material properties don't change). If the thermal properties are not predictable, then the computer model can no longer provide reliable results.

Convection and radiation are typically handled in thermal simulation models by the HTC factor. The values of HTC are reasonably well understood for temperature changes in the range of up to 100°C or so. But they are very difficult to predict if there are hot spots or for a temperature range of 1,000°C, especially if the timing associated with that range of temperature is uncertain.

Dielectric materials, at least from reliable manufacturers, are pretty homogeneous within a manufacturing panel. But they have four parameters of particular interest here. Recall them from Section 2.4:

Tcon: Thermal conductivity is the ability of the material to carry heat away from the source. It can vary widely from material to material. But it can also vary significantly from point to point along a trace.

Tg: Glass transition temperature is the temperature region where the polymer transitions from a hard, glassy material to a soft, rubbery material. This results in dimensional changes and some change in thermal conductivity. This changes the way in which the dielectric is able to conduct heat away from the trace. It is reversible.

Td: The thermal decomposition temperature is the temperature at which weight loss begins to occur. Thermal decomposition is typically specified around 300°C to 350 °C. Thermal decomposition will change the ability of the dielectric to carry heat away from the trace. It is irreversible.

Delamination: In addition, if the board begins to delaminate, all bets are off, and the trace will begin to heat uncontrollably and probably fairly rapidly. The material used in the evaluation to follow had a thermal resistance (time to delamination) specified as

at $T = 260°C > 60$ minutes

at $T = 288°C > 20$ minutes

As temperatures along a trace increase, particularly if they increase quickly, the timing and way these parameters manifest themselves is unpredictable.

12.7 The Fusing Process

The fusing process can be subdivided into two broadly defined categories: strong overload (leading to rapid fusing), and long-term, or slight, overload (leading to slow fusing that might take minutes or hours).

12.7.1 Strong Overload

If we submit a trace to a strong overload (relatively large current), the temperature rises quickly to the point where the weakest spot on the trace melts. The weakest spot on the trace is unpredictable and will typically vary from trace to trace. The melting process will be rapid, particularly because there is positive feedback. That is, the rising current results in rising temperature, which raises the resistance, which increases the temperature, which raises the resistance, and so on.

12.7.2 Slight Overload

If we apply just enough current to ultimately cause the trace to melt, the failure process goes through several separate steps. First the trace temperature will rise to a plateau and seem to level off. If this is near Tg, the temperature will slowly increase until it gets in the vicinity of the delamination temperature. The temperature will start increasing a little faster until it gets near Td. At that point the temperature really accelerates until the trace melts at the weakest spot along the trace. The entire process can take from tens of seconds to hours.

If we plot trace temperature against time, we can't necessarily identify the precise times at which these transitions occur. They will occur at points within, or perhaps underneath, the copper trace itself and by their very nature are point phenomena (remember, the calculated trace temperature is an *average* over the entire trace length, not the temperature at the hottest point).

12.8 Experimental Results

As a result of Prototron's generosity (see Acknowledgments and Section 8.4), there were several boards available for evaluation. The test equipment available included a constant current power supply and a portable oscilloscope. The maximum current capacity of the power supply was 10 amps, which limited the investigation to smaller traces. Separate evaluations suggested the accuracy of the power supply and of the scope were within 2% or 3%.

Traces were 6 inches long and varied in trace width from 15 mils to 27 mils. Most were nominally 0.5-oz foil plated over with 1.0-oz copper, although there was one board that was 0.5-oz foil only. Separate microsection measurements showed that trace widths were well controlled, but the total foil plus plating thickness could vary between 1.9 mils and 2.6 mils, worst case. In this chapter we will report on three cases.

12.8.1 Case A: Fast Fusing

A 15-mil wide, 0.5-oz trace was subjected to a sudden current of 6 amps. The trace fused in 2.75 seconds (see Figure 12.10). The curve of voltage versus time ramps in an almost straight line. Temperature lines are shown on the curve for reference. They are approximate.[5]

A video was taken of the trace fusing. The moment of fusing was really somewhat underwhelming. Fusing occurred within one video frame (1/30 second, see Figure 12.11). There was no smoke and virtually no aroma. There was negligible damage to the board under and around the trace.

A thermal simulation of this trace calculated the fusing time at 4 seconds. This is pretty close and is likely the fusing time had the trace and board been homogeneous. Results are partially predictable because the heating happens so quickly there is little opportunity for any cooling discontinuities to develop. This is evidenced by the lack of any visible damage to the board material. But as discussed above, there are inherent potential hot spots related to the copper and dielectric that can cause localized heating and cooling. The trace fuses at its weakest point, and that point is much hotter than the surrounding trace. There is no way to predict or explain

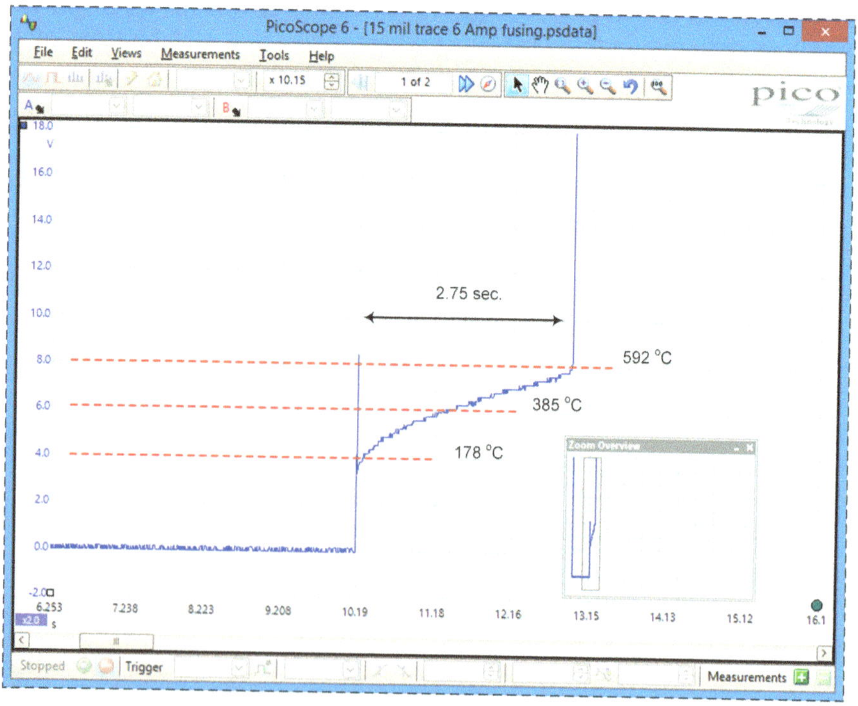

Figure 12.10 15-mil wide, 0.5-oz trace carrying 6 amps.

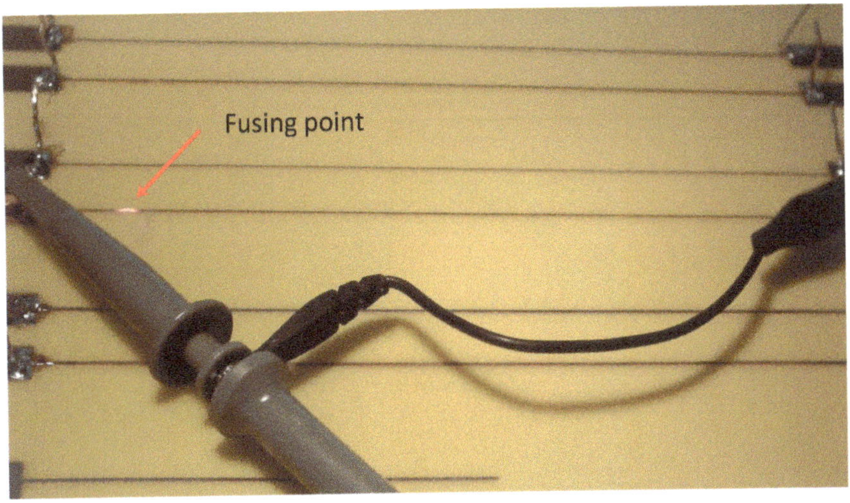

Figure 12.11 15-mil wide, 0.5-oz trace at the moment of fusing.

why that particular point was the weakest point. The *average* temperature of the trace is still pretty low when the trace fuses at its weakest point.

12.8.2 Case B: Slow Fusing

A 20-mil wide, 1.5-oz trace was subjected to 8.5 amps. This trace, being 33% wider and three times thicker than the case above, took much longer, about 30 minutes, to fuse. The curve of voltage (and therefore temperature) versus time is shown in Figure 12.12.

Here the curve tends to level out somewhere around maybe 250 degrees or so and then starts a slow climb upward. As the temperature increases the rate of increase also increases. Some people have described this as a thermal runaway. Once it starts its progress is inevitable (unless the current is interrupted).

When this trace fused the results were much more spectacular. Figure 12.13 illustrates the moment of fusing. First, the trace began to visibly glow about 35 seconds before fusing. It began smoking some 90 seconds before fusing. The aroma of burning material was in the air for perhaps 15 minutes before fusing.

Note in Figure 12.13 how smoke is being visibly ejected under pressure from several places along and under the trace. The board shows visible signs of heat damage, and closer inspection shows continuous areas of what appear to be bubbles (delamination) along the edges of the trace.

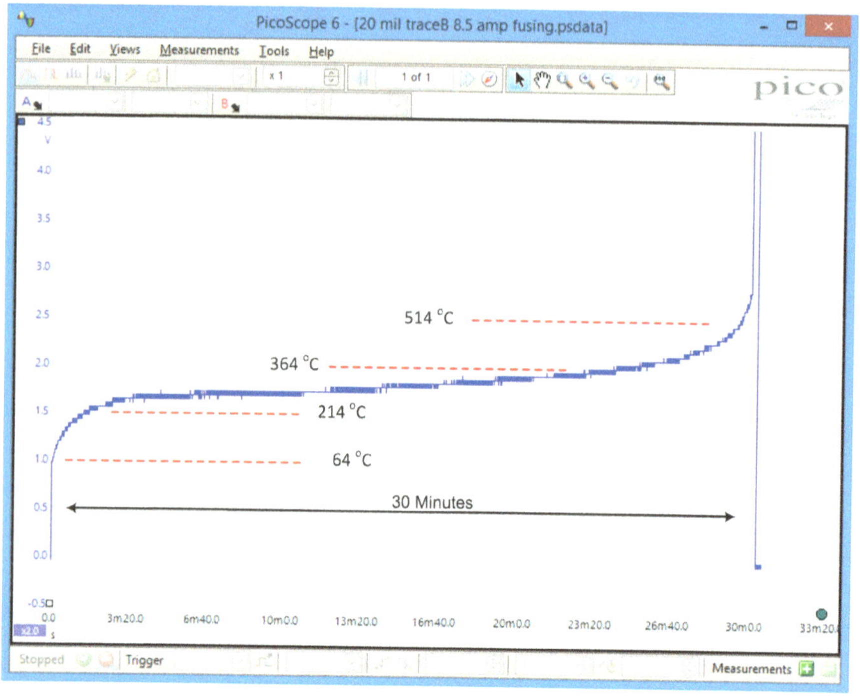

Figure 12.12 20-mil wide, 1.5-oz trace carrying 8.5 amps.

Figure 12.13 Moment of fusing for a 20-mil wide, 1.5-oz trace carrying 8.5 amps.

12.8.3 Other Cases

The current required for the kind of thermal runaway described in Case B, above, is very trace-specific. This is because the discontinuities associated with each trace and each spot on the board are different. For example, Figures 12.14 and 12.15 illustrate the same trace carrying 8.3 and 8.4 amps, respectively. (The 8.4 amps was applied 12 hours after the 8.3 amps, allowing plenty of time for cooling between tests.) At 8.3 amps, the trace showed no signs of running away after 2 hours. The same trace carrying 8.4 amps fused at 1 hour and 16 minutes. Another trace on another board with the same nominal dimensions and carrying 8.5 amps showed no signs of runaway for over an hour, illustrating how there can be minor but important differences between traces and boards.

12.9 Summary

The fusing time of a trace subjected to a heavy current overload (and fusing between 0.0 and maybe 5.0 seconds) is modestly predictable. Under these conditions there is negligible time for any cooling effects to kick in and the

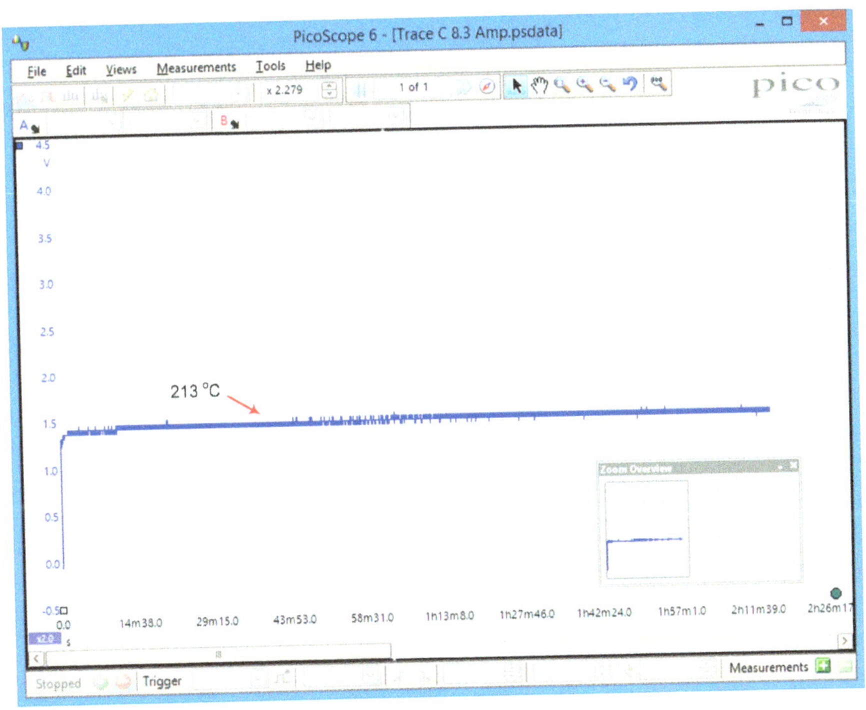

Figure 12.14 20-mil, 1.5-oz trace carrying 8.3 amps showed no signs of runaway after 2 hours.

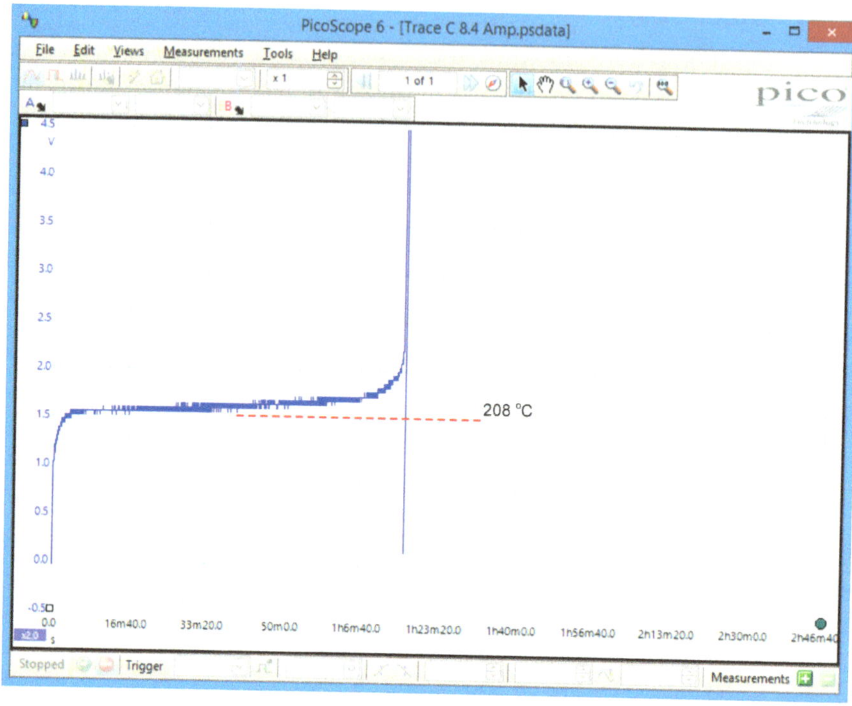

Figure 12.15 The same trace in Figure 12.14 carrying 8.4 amps fused in 1 hour, 16 minutes.

board material properties remain constant. The trace fuses very quickly at a specific point and there is little damage to the surrounding board area. There is little aroma associated with type of failure.

Slight current overloads result in unpredictably long fusing times (as long as hours). This type of overload might occur, for example, if a trace relies on supplemental cooling processes—say flowing air or an embedded heat sink—and that process degrades for any reason. The board material shows significant damage all along the trace, and smoke is apparent. Some smoke is ejected from under the trace under significant pressure. The trace may glow red hot for several seconds before it fuses. There is heavy material damage under and around the trace. When the trace finally fuses, it does so along a small distance (not at a specific point), the copper seeming to curl back as it melts. The aroma of burning is pervasive.

Nominally identical traces have different currents where the slight overload (sometimes referred to as thermal runaway) will result in a fusing condition. This is because of unpredictable discontinuities in the board material and the copper composition.

End Notes

1. A circular mil is the area of a circle with a diameter of one mil. The formula is $A = d^2$. The conversion from circular mils to mil^2 is $\pi/4$. Normal conversions are

   1 mil^2 = 1.273 circular mils

   1 circular mil = .7854 mil^2

   1 m^2 = $19.736*10^8$ circular mil

   1 circular mil = $5.067*10^{-10}$ m^2 = $5.067*10^{-4}$ mm^2

2. From a thermodynamic engineering standpoint, there is a significant difference between the ambient temperature and the initial condition. For example, there could be some current-induced heating of a wire above the ambient temperature in a normal operation before a change of current is applied. In the electronic industry we are less rigorous and tend to refer to the initial temperature of the trace as the ambient temperature. In this book we are assuming that the initial condition (temperature) is the same as the ambient temperature. If the initial temperature of the trace were different from the temperature around the trace, we would refer to the initial trace temperature as the ambient temperature.

3. Preece's equation, Equation (11.2), gives fusing current as a function of area. If we substitute Preece's current (from (11.2)) into (12.2), we can derive an implied time for Preece's equation. It is that value that is plotted on our curves.

4. Furthermore, if we let A = constant, then (12.4) reduces to I^2t = constant. The I^2t value is an important rating in the fuse industry. See http://en.wikipedia.org/wiki/Fuse_%28electrical%29#The_I2t_value.

5. Temperature reference lines in Figures 12.10, 12.12, 12.14, and 12.15 are estimated. The reference resistance (at 20°C) was calculated using nominal trace geometry and a resistivity of 0.67 μohm-in. Resistance at temperature was calculated from V/I (Ohms law). The elevated temperature was then calculated using (12.5) using the thermal coefficient of resistivity as 0.0039. Remember, these are *average* trace temperatures along the trace, not the peak temperature at any point.

CHAPTER

13

Contents

Do Traces Heat Uniformly?

13.1 Bottom Line

- Traces do not heat uniformly along the trace. Nor do they heat uniformly from the midpoint of the trace to either side.
- A significant reason for this nonuniformity is variations in thickness along the trace.

13.2 Background

As discussed in Chapter 12, and shown in some of the following figures, when a trace is heated near the fusing point hot spots develop along the trace. The traces tend to fuse (melt) at a *point* along the trace, not along a broad area, indicating that the traces were not heating evenly at all. We saw similar patterns of uneven heating in other of our fusing investigations. One example is shown in Figure 13.1. It is a thermal image of one of our 20-mil, 0.5-oz. traces heated to about 436°C.

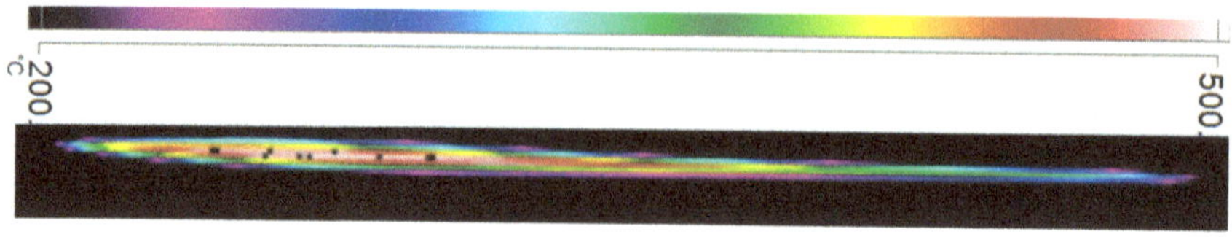

Figure 13.1 20-mil, 0.5-oz trace heated to a maximum of 436°C.

In our testing of trace temperatures using a thermocouple, we noticed that if we moved the thermocouple slightly (right or left) the indicated temperature might change by as much as 1.5°C. This is more than the measurement tolerance (we believed) but still not enough to be able to say whether the observed result was real, or still just an indication of measurement tolerance.

We would expect copper traces to heat relatively smoothly along their length. That is, any temperature gradients would be expected to be smooth. Clearly this expectation breaks down when the trace is heated hot enough. But all the above raises the question: When do hot spots begin to develop along a trace? In particular, do traces, even at low temperatures, exhibit uneven temperatures along their length?

13.3 Thermal Gradients on Traces

In order to answer this question, we needed access to a thermal imager. Scott Dau, a Seattle firefighter and part-time fire investigation instructor, loaned us a Fluke model Ti32 thermal imager. Thermal imagers work on an infrared sensing technology and measure temperatures by measuring the target's infrared characteristics.

Figure 13.2 illustrates the thermal profile of a 100-mil wide, 0.5-oz foil (not plated) trace heated to a maximum temperature of 166°C. It appears about as expected, and is not particularly informative. It is hard to see that the hottest spot is not at the center of the trace, but nearer the left side. This is consistent with where the traces in Chapter 12 fused.

But what is really interesting is if we narrow the thermal range of the image to between 150.0°C and 166.6°C (Figure 13.3). This has the effect of giving us a thermal close-up of the trace. Now it becomes apparent that the heating of the trace is not at all uniform along its length.

But what if the temperature of the trace is raised to a more modest level? Figure 13.4 shows the thermal distribution of the same trace with the maximum temperature raised to only 49.6°C. The nonuniformity of the temperature distribution is evident even over this lower thermal range.

Note also in Figure 13.4 that the trace is cooler along the outside edge than it is down the centerline. This is true in virtually every image. The

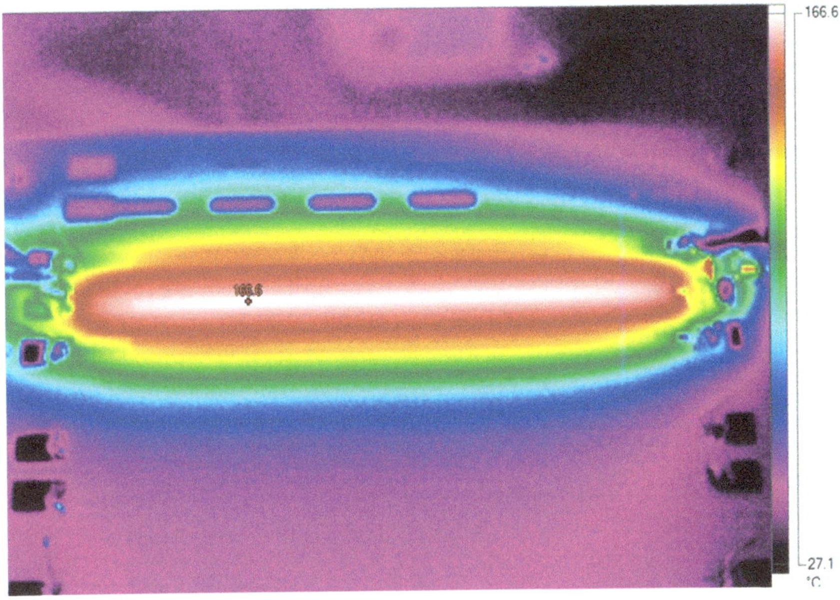

Figure 13.2 Thermal image of 100-mil wide, 0.5-oz trace, heated to about 166°C.

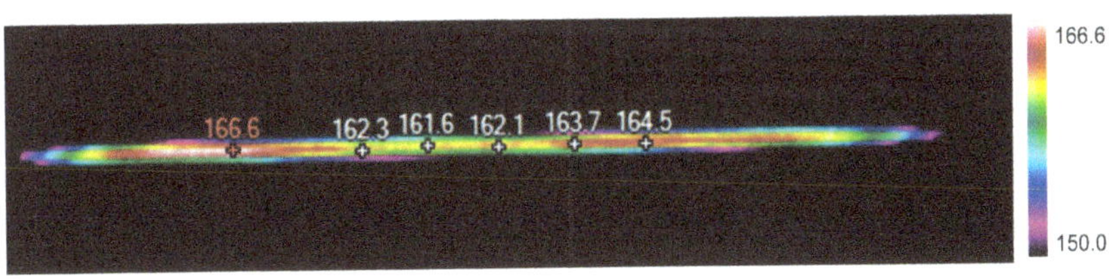

Figure 13.3 Thermal close-up of Figure 13.2.

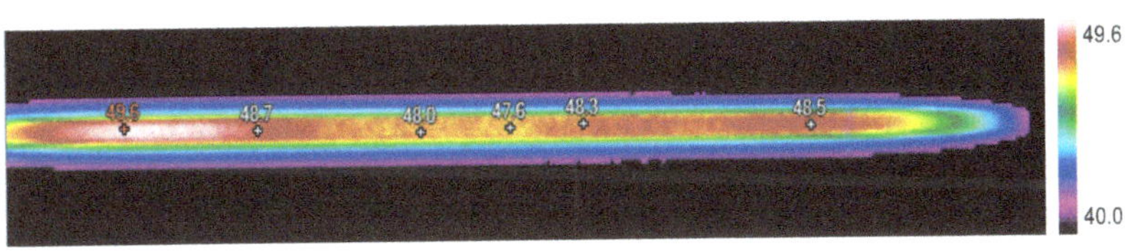

Figure 13.4 Thermal close-up of Figure 13.2 raised to a maximum temperature of only 49.6°C.

reason is that the cooling path from the center of the trace (out to the edge) is longer than it is from the edge of the trace.[1]

13.3.1 Thermal Gradients on Narrow Traces

In the above section we looked at 100-mil wide traces. What about smaller traces? Figure 13.5 illustrates the thermal profile of a 20-mil wide, 0.5-oz trace, raised to a maximum temperature of 62.4°C. It exhibits a nonuniformity very similar to the wider, 100-mil trace.

13.3.2 Does Trace Thickness Matter?

As a result of other investigations, we had a variety of trace thicknesses we could evaluate. They ranged from a simple, 0.5-oz, 100-mil wide trace (nominally 0.65 mils thick) to a complex 2.0-oz, 200 mil wide trace plated up with an additional 2.0 oz of plating, for a nominal total of 5.2 mils thickness. The various combinations are shown in Table 13.1.

The results from these seven options are shown in Appendix I. They all show varying degrees of heating nonuniformity, but we were not able to discern any particular patterns related to trace thickness or degree of copper plating on top of the foil.

13.3.3 Is Trace Thickness Uniform?

During our investigations we microsectioned numerous traces. Figure 13.6 is a microscopic image of one such microsection. It shows a length of trace

Figure 13.5 Thermal profile of a 20-mil wide, 0.5-oz trace raised to a maximum temperature of 62.4°C.

Table 13.1
Trace Options Available for Thermal Hot-Spot Analysis

Width (mil)	Foil (oz)	Plating (oz)	Total (Nominal) Thickness (mil)	Current Through Trace (A)	Maximum Temperature (°C)
100	0.5	—	0.65	5.6	53.0
100	0.5	1.0	1.9	8.0	54.0
100	1.5	—	2.0	7.5	54.5
100	1.5	1.75	4.4	10.0	47.0
200	2.0	—	2.6	10.0	39.0
200	2.0	1.0	3.9	10.0	35.0
200	2.0	2.0	5.2	10.0	33.6

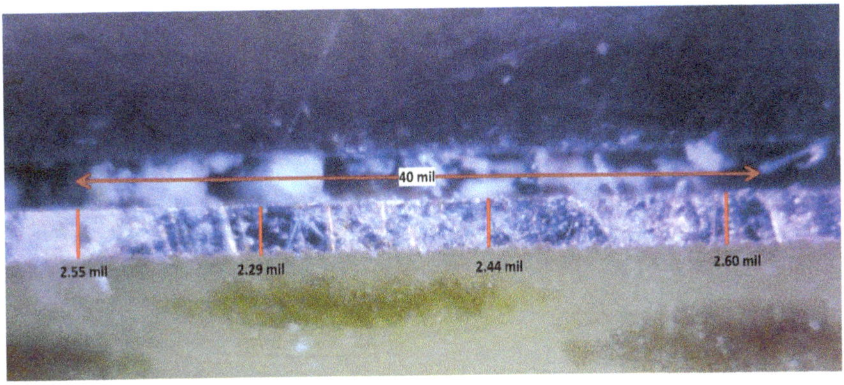

Figure 13.6 Trace thickness is not uniform along a trace.

a little over 40 mils in length. Within that section, there is a difference in trace thickness varying from 2.60 mils to 2.29 mils, a difference of 0.31 mils. Recall that in Section 7.2.5 we showed a case where the temperature changed 1.0°C for every 0.03-mil change in thickness. We have also mentioned earlier that copper plating during the board manufacturing process can be quite variable in thickness over various parts of a board. The bottom line is that trace thickness is definitely not uniform from point to point on the board.

13.3.4 What Causes Thermal Nonuniformity?

At this point it is unknown exactly what causes the thermal nonuniformity in the heating of the traces. There are a variety of possibilities:

- Contamination under the foil or under the copper plating (if any).
- Contaminants in the copper alloy itself.
- Variations in trace thickness or width due to normal manufacturing tolerances.
- Variations in the properties of the underlying dielectric material. Note how part of the trace in Figure 13.6 is directly over fiberglass and part of it is directly over resin. We mentioned in Section 4.4.1 that we believe these two areas would have different cooling characteristics and therefore different temperatures.

13.3.5 Conclusion

It is clear that traces do not heat uniformly once high levels of temperature are reached. The fact that traces at high temperature fail at random points is clear evidence of this.

It also seems apparent that traces do not heat uniformly even at lower temperatures. It can be stated with some confidence that traces are hotter down the centerline and they cool toward the edges, primarily because the cooling path is shorter the nearer one is to the edge.

13.4 Thermal Gradients Around Corners

Through the years the impact of right-angle corners on PCB traces has been a hot topic. As early as the 1990s some people were arguing strongly against the use of 90° corners, using mitered corners instead. They based this recommendation, typically, on one of two reasons:

1. The width of a typical PCB trace increases as the trace turns a corner. This increases the local capacitance at that point, lowering the trace impedance. This impedance discontinuity can therefore set up a reflection that can lead to signal integrity issues.
2. The current density around a trace corner shifts so that it is strongest at the inside edge of the corner where the path length is the shortest. This increased current density can lead to an increase of the local electromagnetic field, which can subsequently lead to increased electromagnetic interference (EMI) from the trace.

In a landmark paper published in 1998,[2] Brooks tried to put these arguments to rest. In that paper he showed:

- Although it is true the width increases, it does not increase very much. Furthermore, the maximum width occurs *at a point*, not over a range. And finally, the entire signal path around the corner is *much* shorter than the critical length.[3] Therefore, the impedance discontinuity is insignificant.
- An empirical investigation of actual EMI measured from traces with various corner designs, including a diabolical 135-degree corner with sharp points (which would never be found on a PCB!) showed no significant radiation greater than a control straight trace.[4]

But one question remains: Are there thermal issues related to right-angle corners? This chapter looks at the thermal gradients around the corners on a PCB.

13.4.1 Software Simulation

We used TRM to look at traces with right-angle corners. Four corner configurations were looked at, illustrated in Figure 13.7. Configuration (a) is a constant-width corner (i.e., the trace is the same width everywhere). It has a rounded inner corner. Configuration (b) is typical of a trace found on a PCB. It has mitered 45-degree corners whose outer diameter is the same as the trace width. Configuration (c) is also typical of what you might see on a PCB. It is typically created by drawing a trace with a circle (whose diameter is the same as the trace width). It has a rounded outer corner with a sharp, 90° inner corner. Configuration (d) is a square corner, rarely seen on a PCB, and actually somewhat difficult to create on a PCB (it can be drawn with a square aperture on some plotters). It should be noted that the width across the corner, and therefore the trace cross-sectional area at the corner, is the same as the body of the trace for configuration (a), but increases as we move to configuration (b), (c), and (d).

The basic model we used was a 5-mm (200-mil) wide, 35-μm (1.0-oz) trace on a standard 1.5-mm thick FR4 substrate. All four trace configurations were modeled and driven with a current of 8.0A, resulting in a temperature of approximate 44°C (about a 24°C increase). The TRM model allows us to investigate (among other things) both the current density and the thermal characteristics around the corner.

Figure 13.8 shows the current density (A/mm^2) around the corners of the four configurations. There is a fairly dramatic shift in current from the outside corner to the inside corner, especially for configurations (c) and (d), where the inside corners are sharp.[5] This directly relates to the path length of the current around the corner, that path being significantly shorter along the inside edge.

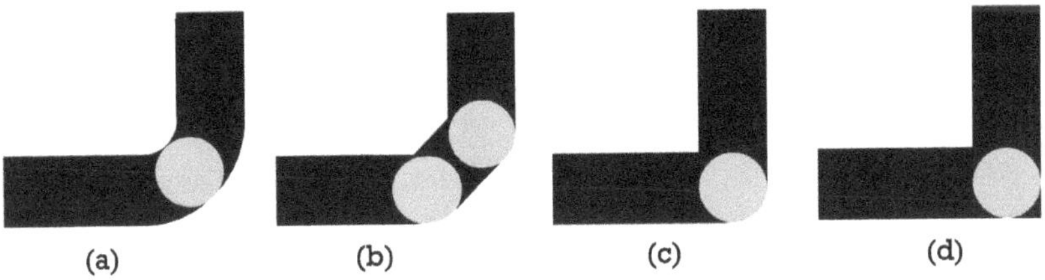

Figure 13.7 Four different corner configurations.

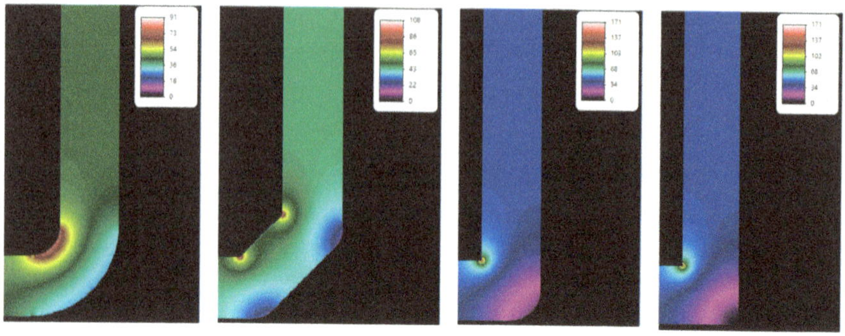

(a) (b) (c) (d)

Figure 13.8 Current density around the corners of the four trace configurations shown in Figure 13.7, each carrying 8.0A.

One might speculate that the increased current density at the inside edge of the corner would lead to a temperature increase along the trace at that spot. Another argument, however, might be that the thermal conductivity of the copper is so high that no temperature gradients could develop. Figure 13.9 shows the thermal profiles for the four configurations modeled. The small boxes record the temperature at the upper left corner of the box.

The models suggest that the trace temperature is slightly warmer at the inside edge of the configurations than at the outside edge. The temperature differences range from about 0.5°C to about 1.5°C, and are larger the greater the width at the corner exceeds the width of the trace. Interestingly, however, the inside corner is either about the same temperature as the basic trace or is cooler than the basic trace. This is because the increased width at

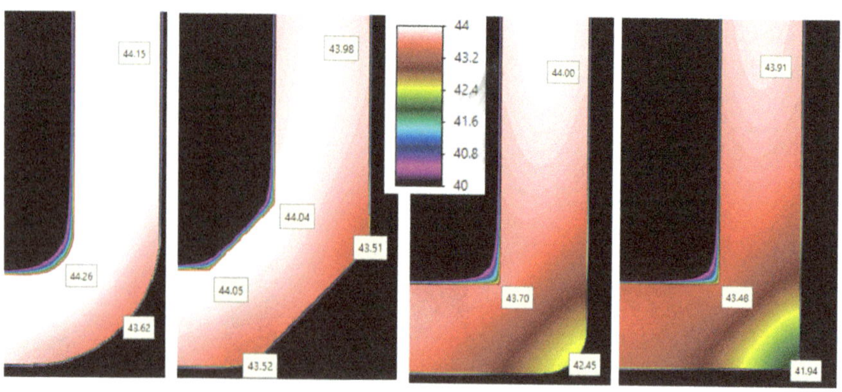

(a) (b) (c) (d)

Figure 13.9 Thermal profile simulation around the corners of the four trace configurations. The temperature range shown is between about 40°C to 44°C.

the corner offers a lower resistance to the current through the corner and therefore a lower I^2R heating around the corner.

But it turns out the heating dynamics around the corners are much more complex than this. Figure 13.10 illustrates the thermal profiles of trace configurations (a) and (c) (the constant width corner and the drawn corner) and of the board around the traces. Note that the board around the inside of the corners has a wider thermal profile than around the outside of the corner. This indicates the outside of the corner is cooling more efficiently than is the inside of the corner. There are at least four factors that might contribute to the thermal profile of the trace and the surrounding area:

A. If the cross-sectional area of the corner is larger than the cross-sectional area of the rest of the trace, the point resistance around the corner will be lower than the point resistance along the rest of the trace. Therefore, the temperatures around the corner will tend to be lower than they are along the rest of the trace.

B. The current density is significantly higher toward the inside of the corner than it is toward the outside of the corner. Therefore, the point I^2R power dissipation will be higher at the inside of the corner than it is at the outside of the corner.

C. The thermal conductivity of the copper trace is so high that some of the heat generated at the inside corner will conduct away through the copper to the outside corner.

D. But notice how the *board area* at the inside of the corner tends to be much warmer than the board area at the outside of the corner. This is because the heat from both legs of the trace conducts into

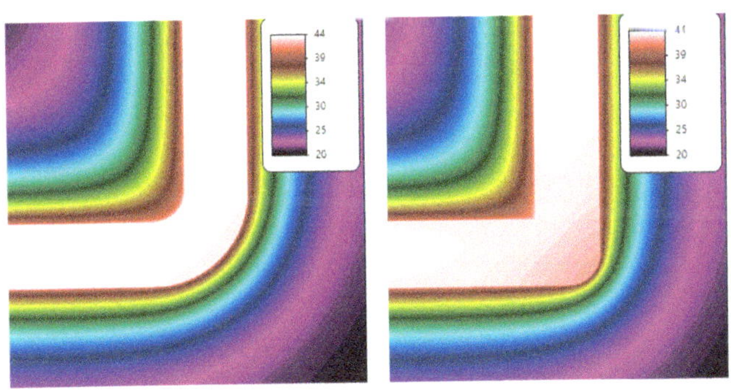

Constant-Width Corner Drawn Corner

Figure 13.10 Thermal profile of the modeled traces and the board area around the traces.

the same 90° arc of the board on the inside of the corner, but the heat tends to fan out (in a 270° arc) away from the outside of the corner. Thus, the outside of the corner cools much more effectively than does the inside of the corner. This is the most important determinant of the trace temperature around the corner.

Now, looking back at Figure 13.10, configuration C (Drawn Corner), it is true that the inside of the corner is warmer than the outside of the corner, it is *also* true that the inside edge is *cooler* than the midpoint of the straight section of the trace on either end of the corner. This illustrates the impact of the complex relationships outlined in A through D above.

Additional modeling, not reported here, results in the following observations:

1. The greater the distance from the inside corner to the outside edge of the curve, the more pronounced is the difference in temperature;
2. The sharper the inside corner, the more pronounced is the difference in temperature (e.g., smooth inside curves result in lower temperature differences around the corner);
3. Narrower traces and/or lower currents result in smaller temperature differences around the corner.

13.4.2 Experimental Verification

The experimental investigation was performed by Norocel Codreanu[6] with a FLIR SC640 infrared (IR) camera. The electromagnetic radiation in the 7.5–13 μm range hits the active layer in the camera and changes its electrical resistance. This resistance change is measured, processed, and transformed by the IR camera into a temperature array that can be represented graphically.

The test board consisted of a standard FR-4 substrate, 1.5 mm (60 mil) thick with a copper thickness of 35μm (1.0 oz). The traces were 5.0 mm (200 mils) in width. The traces of interest were drawn with corner configurations (c) and (d) above. They were driven with 8.0 amps of current.

The traces were then photographed with the thermal imager. Figure 13.11 shows the thermal pattern for the drawn corner (left) and the square corner (right). Compare these results with those shown in Figure 13.9 configurations c and d. The thermal images compare favorably with the modeled results.

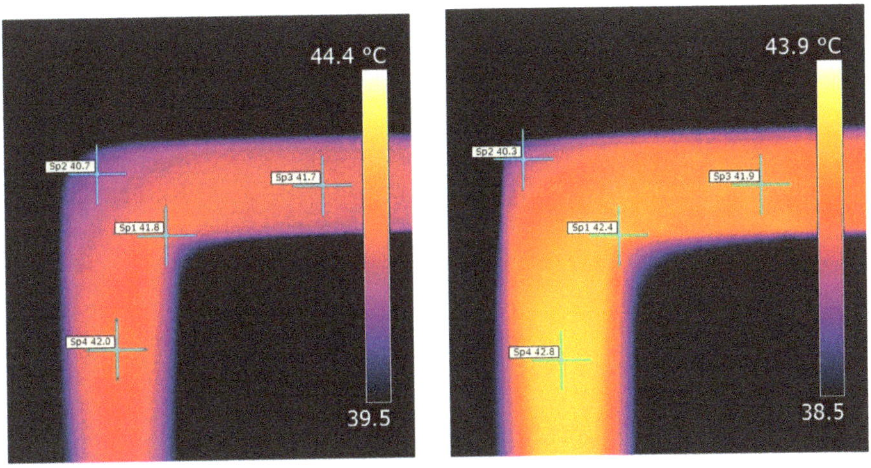

Figure 13.11 Thermal profiles of the test board: drawn corner (left) and square corner (right).

Figure 13.12 illustrates the thermal profile of the same traces as shown in Figure 13.11 along with the surrounding board area. Compare these images with those shown in Figure 13.10. Again, the two results compare favorably. This illustrates how the outside of the traces cool more effectively than do the inside of the traces.

13.4.3 Conclusions

Current densities around corners are higher at the inside of the corner compared to the outside of the corner. This is consistent with previous studies.

Temperatures are higher at the inside of a corner than they are at the outside of a corner. And there is a close correlation between the modeled

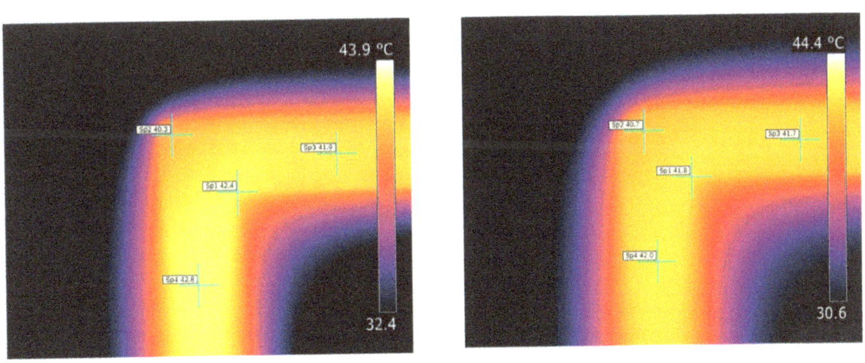

Figure 13.12 Thermal profiles of the test board and surrounding board area: drawn corner (left) and square corner (right).

data and the experimental data. But these temperatures are not necessarily higher than they are along the main body of the trace. This is because the traces are wider at the corner than they are along the main body, and the electrical resistance is therefore lower. Consequently, the I^2R heating is lower at that point. Furthermore, the outside edge of the corner cools more efficiently than does the inside edge, further contributing to the temperature imbalance.

However, the final conclusion is that the temperature imbalances around a corner are not the result of the difference in current densities. They are the result of the significantly more efficient cooling mechanism around the outside of the corners where the open board area is much wider.

End Notes

1. See Section 12.5.2 for a discussion about this.
2. Brooks, D. G., "90 Degree Corners: The Final Turn," *Printed Circuit Design Magazine,* January 1998, available at http://www.ultracad.com.
3. For a discussion of critical length, see Brooks, D., *PCB Currents; How They Flow, How They React,* Prentice Hall, 2013, p. 216.
4. While it is true that there are no problems with using right-angle corners on traces, a great many designers still prefer mitered corners for aesthetic reasons. Some fabricators prefer mitered corners to right-angle corners for fabrication reasons.
5. If a model has a sharp, 90-degree corner, the mathematics may result in an infinite current density at that specific (infinitely small) point because of a singularity (think divide by zero). This does not have any effect on the rest of the model.
6. We are indebted to Norocel Codreanu, Ph.D., who conducted the experiments and provided these images. Dr. Codreanu is full professor at Politehnica University of Bucharest (UPB), Romania, Faculty of Electronics, Telecommunications and Information Technology, and currently the executive manager of the Center for Technological Electronics and Interconnection Techniques (UPB-CETTI), a UPB research center.

CHAPTER

14

Contents

Stop Thinking about Current Density

14.1 Bottom Line

- Stop thinking in terms of current density! There is little or no correlation whatsoever between current density and trace or via temperature.

14.2 Background

Many design engineers and even many software suppliers make the significant mistake of equating changes in trace or via temperature with current density. This is incorrect at best and dangerous at worst. There is little if any correlation between temperature and current density. Current and trace dimensions (among other things) are the relevant variables, but current density is not. We hope by the end of this chapter you will understand why. Here are some illustrations that will help you understand this.

14.3 Current Density Is Not an Independent Variable

Current density is a *derived* variable. It is current *divided by* trace cross-sectional area, or C/(width * thickness). When we say that we need a trace width that can handle a current density of *J*, we implicitly already knew the width in calculating the current density in the first place. The width (or area) cannot be a constant on one side of the equation (current density) and an unknown variable on the other side of the equation.

We can also see this looking at the formulas we derived in Section 5.2.2. Here is (5.5) repeated as (14.1):

$$\Delta T = 215.3 * C^2 * W^{-1.15} * Th^{-1.0} \tag{14.1}$$

Since current density, *J*, is current divided by width*thickness, we can rewrite this equation as follows:

$$\Delta T = 215.3 * J^2 * W^{0.85} * Th \tag{14.2}$$

We cannot plug values for width and thickness into this equation without adjusting *J* by the same amounts. So it just doesn't make sense to say that temperature is a function of current *density*.

14.4 IPC Curves

It is fairly easy to demonstrate from the IPC curves themselves that there is not a direct relationship between temperature and current density. The black lines in Figure 14.1 are derived from Figure A-26 on page 42 in IPC-2152. They are for 2.0-oz external conductors. The horizontal axis is trace cross-sectional area, the vertical axis is current, and the black curves are constant temperature curves.

Now current density is current divided by cross-sectional area. So current density is a straight line on these axes. Four different values of current density are drawn (in red) on these axes. Every place a red line crosses a black curve represents a different trace current/temperature relationship but with the same current density.

14.5 Copper Type

Suppose we have two identical traces; they have the same width and thickness and they carry the same current. Therefore, they have the same current density. But one trace is formed from deposited copper and the other

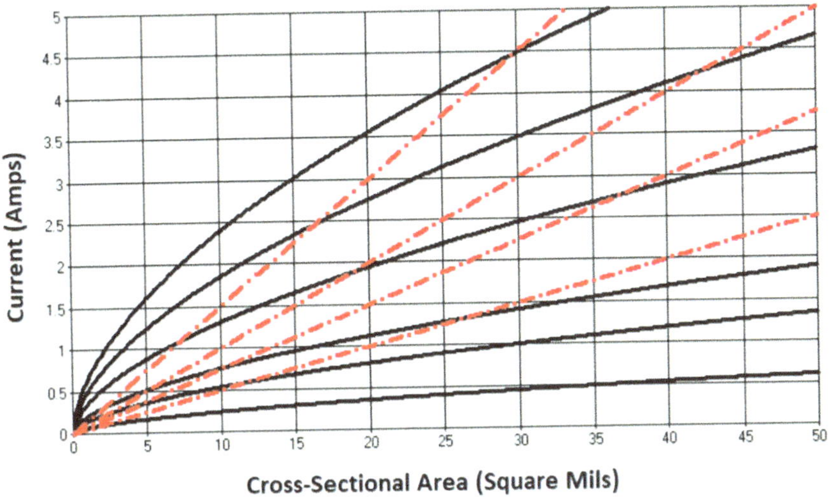

Figure 14.1 Constant current density lines (red) intersect several different temperature curves on the IPC-2152 charts.

from rolled copper. The resistivities of the copper on these traces is nominally 1.64 and 1.76 μohm-cm, respectively. The temperature of the traces is directly related to resistivity, so the rolled copper trace will be some 7.3% hotter than the deposited copper trace, even though they have the same current density.

14.6 Dielectric Type

Again, suppose we have two identical traces; they have the same width and thickness, made from the same copper type, and they carry the same current. Therefore, they have the same current density. But one is over polyimide and the other is over FR4. They will have different temperatures because of the cooling characteristics of the different dielectrics, despite the fact they have the same current densities.

14.7 Right-Angle Corners

However, as we concluded in Section 13.4.3, temperature imbalances are the result of different cooling mechanisms, not the result in differences in current densities.

14.8 Trace Form Factor

Traces having the same cross-sectional area and carrying the same current (and therefore having the same current density) will have different temperatures depending on their form factor. For example, thinner, wider traces will be cooler than narrower, thicker traces, because thinner traces have more surface area in contact with the dielectric and cool more effectively.

We can easily verify this point using (5.5) we developed in Section 5.2.2. Table 14.1 was developed using that equation. It shows that traces with the same cross-sectional area and carrying the same current increase in temperature as the thickness increases and the width correspondingly decreases.

14.9 Via Current Densities

One of our biggest concerns is when designers base decisions regarding the number of, or the size of, vias on the current density within them. We have already shown (Sections 8.3.7 and 8.5.2) that the temperature of a via is not related to the magnitude of the current flowing through it (and therefore, not the current density). It depends on the temperature of the parent trace. Recall that is because the trace acts as a huge heat sink for the via and conducts the heat away from the via.

We recently ran another set of via simulations regarding current density. The parameters of the simulations are as follows:

Board size: 80 × 12 mm

Board thickness: 1.6 mm

Copper thickness: 0.04 mm

Table 14.1
Trace Temperature Is Not a Function of Current Density

Oz	Thick (mil)	Width (mil)	Area (mil^2)	Current (amp)	Current Density (A/mil^2)	Trace Temperature (°C)
0.5	0.65	600	390	15	0.0385	47.5
1.0	1.3	300	390	15	0.0385	52.8
1.5	1.95	200	390	15	0.0385	56.1
2.0	2.6	150	390	15	0.0385	58.6
3.0	3.9	100	390	15	0.0385	62.3
4.0	5.2	75	390	15	0.0385	65.0
5.0	6.5	60	390	15	0.0385	67.2

Trace length: 32 mm, top and bottom; 60 mm overall

Trace width: varies; 0.6 to 5.0 mm

Via diameter: 0.26 mm

Via wall thickness: 0.04 mm

Via cross-sectional area: 0.02675 mm^2

The trace width was varied from 0.6 to 5.0 mm, but the via dimensions *remained constant* for all simulations. The applied current was adjusted so that the via temperature also remained constant (to the degree practical). The trace temperature for each simulation was recorded at a point midway between the beginning of the trace and the center of the via. Figure 14.2 illustrates the thermal pattern for the top layer of the simulation for the 2.0-mm trace carrying 5.25A.

The results of the simulations are shown in Table 14.2 and Figure 14.3. The first thing to note is that when the trace width is small, such that the via cross-sectional area is about the same as or larger than that of the trace, the via temperature is *less than* the trace temperature. Recall this is because the via resembles an internal trace, completely surrounded by board dielectric, and so it cools more efficiently that does the parent trace. As currents (and

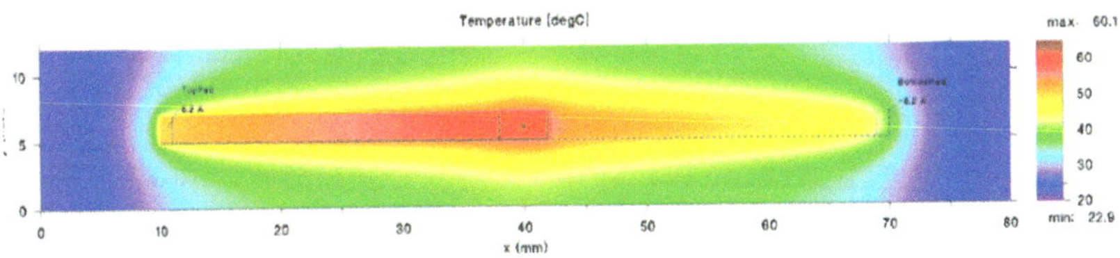

Figure 14.2 Thermal profile of the 2.0-mm trace carrying 5.25A. Via is at the center of the board.

Table 14.2

Summary of the Constant Temperature Via Simulation

Trace W (mm)	Current (A)	T-Trace (°C)	T-Via (°C)	J-Trace (A/mm^2)	J-Via (A/mm^2)
0.6	2.74	61.2	60.4	114.2	99.1
0.7	2.95	59.5	59.4	105.4	106.7
1.0	3.57	58.6	60.5	89.3	129.1
2.0	5.25	56.5	60.5	65.6	189.9
3.0	6.5	54.8	59.8	54.2	235.1
4.0	7.5	53.7	60.0	46.9	271.2
5.0	8.4	52.7	59.6	42.0	303.8

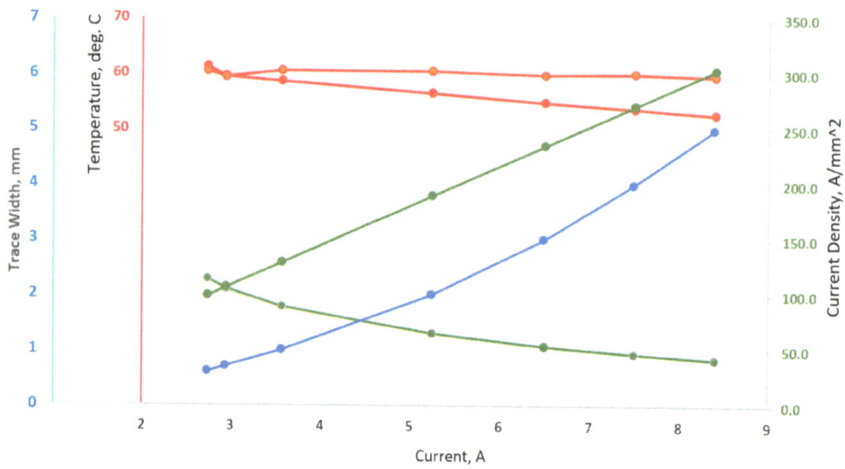

Figure 14.3 Constant-temperature via simulation.

therefore temperatures) increase, vias do get warmer than the parent trace, but only slightly so.

Second, note that even though the via dimensions remain constant, the temperature is remaining (nearly) constant over the entire current range from 2.7 to 8.4 amps. That is because the parent trace is acting as a heat sink. Without the effects of the parent trace, the via would likely melt at 8.4 amps in approximately 1 second.

Finally, and most important, note that the current density (green curves) is varying widely, both along the trace and through the via, as we change simulations. Yet the via temperature is remaining (mostly) constant, and the trace temperature is only varying over a narrow range. This shows that current density is not related to via temperature.

14.10 Conclusion

It bears repeating: Stop thinking in terms of current density! There is almost no situation where thinking in terms of current density will lead to a correct solution. And it might lead you to a very incorrect solution. There is little or no relationship between current density and trace or via temperature.

CHAPTER

15

Contents

AC Currents

15.1 Bottom Line

- The temperature of a trace is determined by the equivalent RMS value of the current driving it.
- The temperature of a trace at any frequency and duty cycle seems to be independent of frequency (at least at frequencies below where the skin effect comes into play).

15.2 Digital Simulation Models

In previous chapters, all models and experimental results were based on switched, constant (DC) currents. In this chapter we will look at what happens when we apply various types of AC currents to a trace.

One relatively simple way to utilize TRM to evaluate AC current is to construct a basic model of a trace and apply the appropriate (adjusted) current to it. The trace model we have chosen is the same trace we used in previous simulations (Section 8.3.3), a 26-mil wide,

2.9-mil thick trace on a dielectric with known thermal conductivity (0.7 W/mK), mass density (2,000 kg/m³) and specific heat capacity (900 J/kg-K). We will apply 5.0 amps DC to the trace as a reference. We will then assume the applied current is a square pulse with a varying duty cycle. The appropriate equivalent DC current for each duty cycle is the effective RMS current (for a square wave), given by (15.1):

$$\text{RMS} = 5 * \text{SQRT}(\text{DCy}) \tag{15.1}$$

where DCy is the duty cycle expressed as a fraction. Table 15.1 provides the detailed equivalent currents and the maximum temperatures from the TRM model for that particular equivalent RMS current. (These data calculations include the effects of the thermal coefficient of resistivity.)

The result is a linear curve as shown in Figure 15.1. (The slight glitches in the curve are caused by adjustments to the HTC value in the model. See Section 6.3. The curve should be perfectly linear.) The maximum trace temperature is 65.7°C, with 100% duty cycle (i.e., direct current). It declines in a linear fashion to the ambient room temperature at a duty cycle of zero (i.e., no current). The ambient temperature is set to 26°C in this model for reasons that will be clearer below.

Figure 15.2 illustrates a graph of temperature versus time for the 5.0-amp DC current (black line) and for 3.536-amp DC (dotted red line), which is the RMS equivalent current for the 5.0-amp current at a 50% duty cycle.

Table 15.1
Equivalent RMS Currents and
TRM Temperature Calculation

Duty Cycle (%)	RMS Current (A)	Maximum Temperature (°C)
100	5.000	65.7
99	4.975	65.3
90	4.743	61.2
80	4.472	56.8
70	4.183	52.6
60	3.873	49.5
50	3.536	45.4
40	3.162	41.2
30	2.739	37.3
20	2.236	33.8
10	1.581	29.8
1.0	0.50	26.0

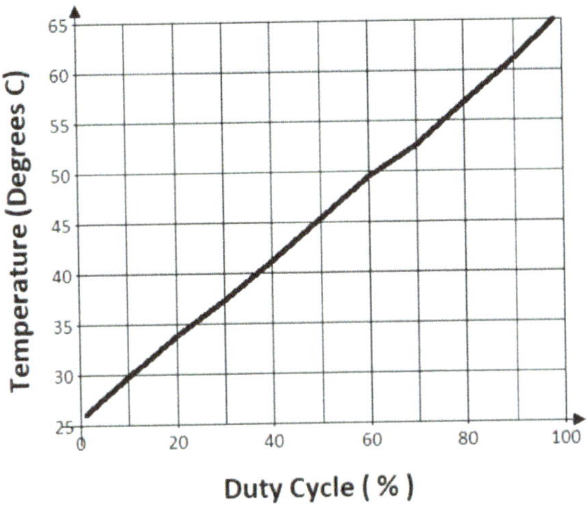

Figure 15.1 Trace temperature in our model from a 5.0-amp pulse as a function of duty cycle.

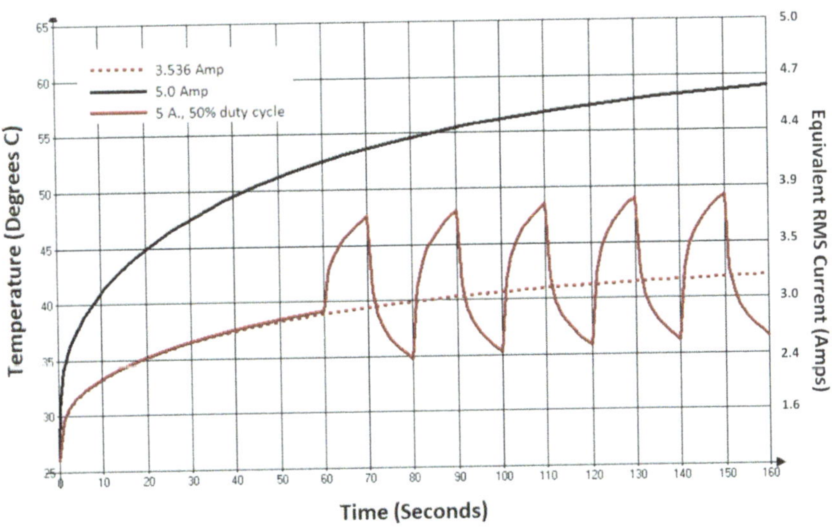

Figure 15.2 Temperature curves for a 5.0-amp, 0.05-Hz waveform at 50% duty cycle (red) along with 5.0-and 3.536-amp curves.

These curves will rise asymptotically to the expected values of 65.7°C and 45.4°C, respectively, as expected. (Recall it can take up to 5 or 6 minutes for trace temperatures to stabilize. See Section 7.2.2.) The scale on the right-hand axis is the equivalent RMS current (amps) that will lead in the long term to the corresponding temperature on the left-hand axis.

TRM offers an alternative way to look at an alternating current waveform. That is to apply current to the waveform according to a cyclical schedule supplied by the user. The schedule is provided by way of a data series entered into an Excel spreadsheet. An example is shown in Table 15.2. Time is the time in seconds. "Power" and "Return" are the current input and output pads, respectively, for the model. The values in the table are the current that will be applied to those pads until the next time specified. This example is for a 5.0-amp current (RMS value is applied for the first 60 seconds in order to speed up the heating, and then at a 0.05-Hz frequency after the first 60 seconds) with a 50% duty cycle.

The solid red curve in Figure 15.2 is the resulting waveform. Its mean temperature is the same as that found in Figure 15.1, trending to about 45.4°C.

TRM offers yet a third way to look at all this. First, we run a transient model at 5.0 amps (the black curve in Figure 15.2), which derives a heating curve. Then we use the pulse width modulation (PWM) option found under the "Extra" menu (Figure 15.3). This modulates the heating curve with a duty cycle. Setting the "Time interval t2" and "Time interval t3" each to 10 seconds gives us a 20-second cycle or 0.05 Hz. The 1.464-watt power value comes from a basic model run of 5.0 amps. It is the power dissipated in the trace (= I^2*R).

Figure 15.4 is the resulting waveform. The shape of this waveform looks very similar to the waveform in Figure 15.2. (Technical note: The PWM approach will slightly understate the other approaches in amplitude because it does not yet take the thermal coefficient of resistivity into consideration

Table 15.2
Cyclical Input Data for TRM Analysis

Time	Power	Return
0	3.5355	–3.5355
60	5	–5
70	0	0
80	5	–5
90	0	0
—	—	—
150	0	0
160	5	–5

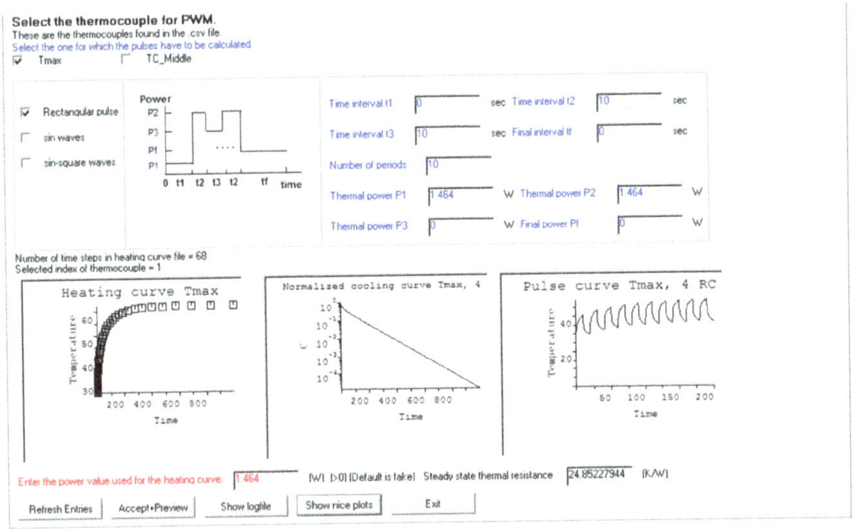

Figure 15.3 Setting up the PWM function in TRM.

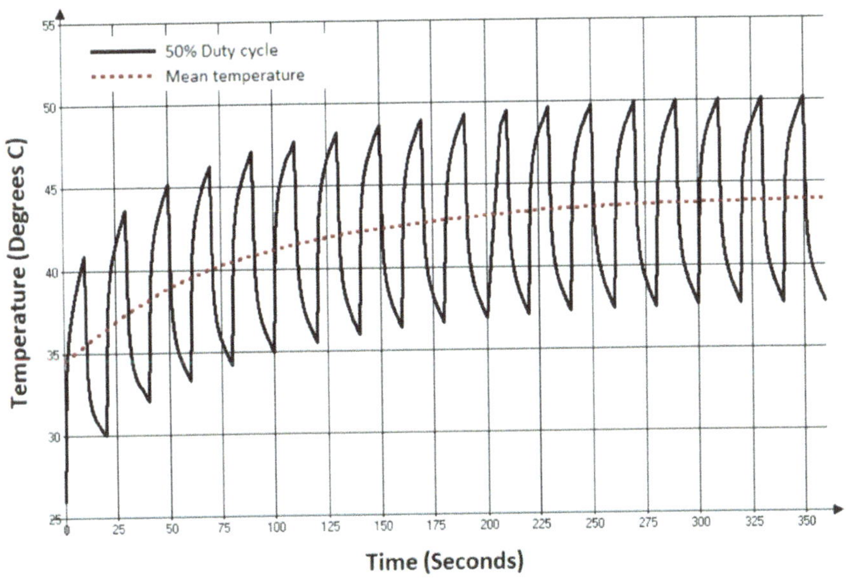

Figure 15.4 The resulting waveform calculation from the PWM function.

for the calculations.) Therefore, these latter two approaches to determining the waveform of an AC signal are equivalent, but this last approach is the easiest and most versatile, and requires minimal CPU time.

This now gives us an easy way to look at other duty cycles. For example, let's set t2 and t3 to 5 and 20 seconds, respectively (and then vice versa). This gives us a frequency of 0.04 Hz, at a 20% or 80% duty cycle. This results in the curves shown in Figure 15.5. TRM's PWD function also provides the mean temperature for each cycle in this waveform. The trend line for the mean values approaches the 33.8°C maximum temperature for the 20% duty cycle and the 56.8°C maximum temperature for the 80% duty cycle provided in Table 15.1.

So far, the TRM models seem to make some intuitive sense at lower frequencies. The PWM function will also enable us to estimate the temperatures of traces at higher frequencies. Figure 15.6 is the result from the PWM function at 100-Hz, 50% duty cycle. The cycles are now so short that the individual cycles are not visible, and have no effect on the result.[1] The curve looks like the curve for a DC current at the equivalent RMS level (3.54 amps) and levels off at about 44°C. This equates to a value from Figure 15.1 of 45.4°C (but without the thermal coefficient of resistivity adjustment).

15.2.1 Preliminary Results

The conclusions so far would seem to suggest:

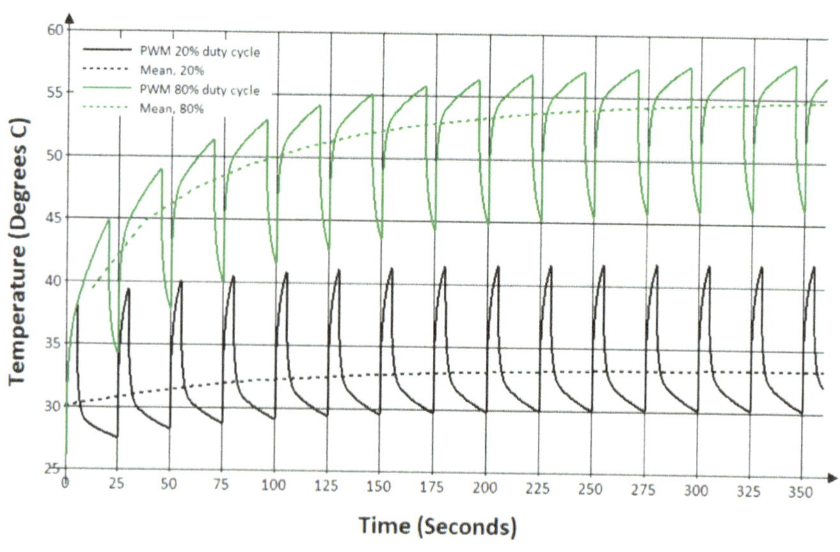

Figure 15.5 The resulting waveform calculation from PWM for a 0.04-Hz frequency, and 20% and 80% duty cycle waveforms.

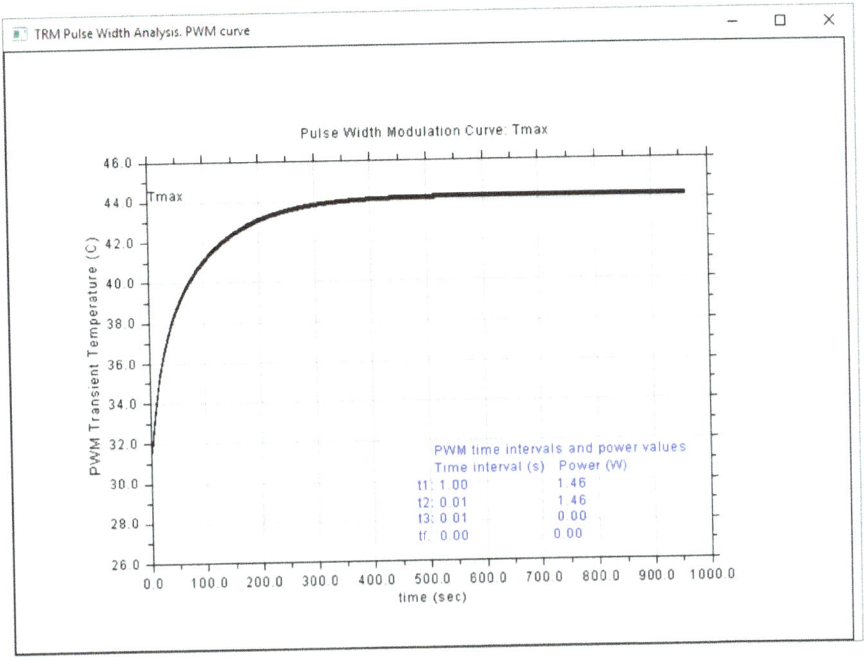

Figure 15.6 PWM result for a 100-Hz, 50% duty cycle input current.

- The temperature of a trace is determined by the equivalent RMS value of the current driving it;
- The temperature of a trace at any duty cycle seems to be independent of frequency (at least at frequencies below where the skin effect comes into play).

15.3 Experimental Verification

We had some boards left over from our other investigations to use in our experimental testing. One of the traces on each board is identical to the model used above. However, our laboratory facilities and equipment were severely restricted. We did not have the equipment available for adequate testing. We describe here what we were able to accomplish with limited resources and what the results suggest to us. The diagram of the test procedure is shown in Figure 15.7.

A constant current generator is set to 5 amps. It is important that the current flow not be interrupted, otherwise the generator output will change dramatically, trying to maintain the constant current flow. Therefore, it is important to maintain a constant current from the generator, even though

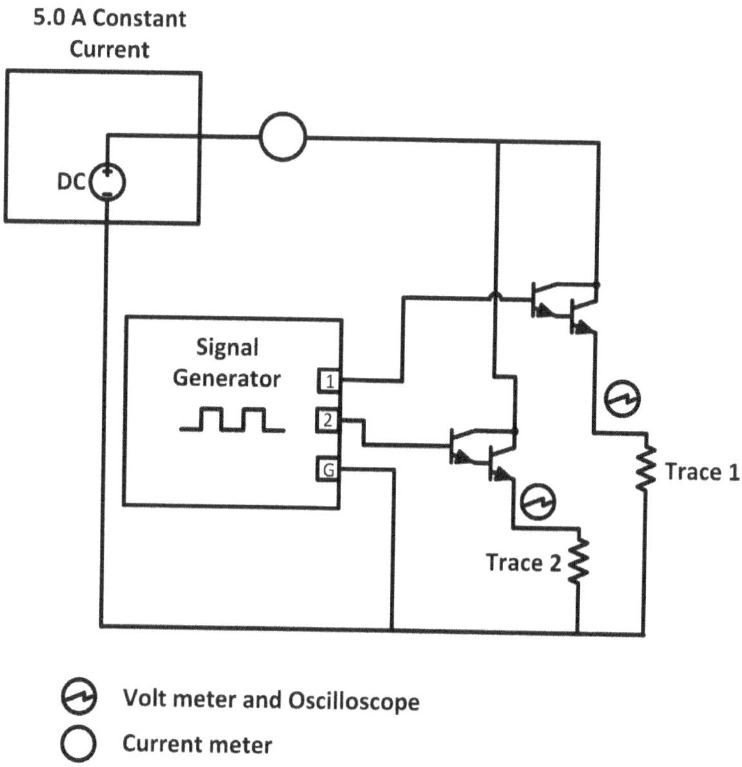

Figure 15.7 Diagram of experimental test setup.

the current through a trace is changing. To achieve this, we used two identical test traces with current switching between them very quickly. The current was switched by a pair of Darlington transistors, each driven by a waveform generator. The waveform generator has two complementary outputs. These outputs were set to a square wave with output between 0.0 and 4.0 volts. The frequency was adjustable between 0.01 Hz and over 20.0 MHz in 0.01-Hz steps. The duty cycle could be set from 0.1 % to 99.9% in 0.1% steps. Current and voltage meters, and oscilloscope probes, were set where indicated. The temperature of each trace was measured with a thermocouple data logger with a sampling rate of 1.0 second and a resolution of 0.5°C. The ambient environment was a typical office environment with temperature about 26°C.

For the first test, the duty cycle was set to 50% and the frequency varied from 0.05 to 10,000 Hz. The results are shown in Figure 15.8. The temperature is (approximately) constant at 45°C within the resolution of the thermocouple. This compares to the 45.4°C temperature recorded in Table 15.1 and is consistent with expected results.

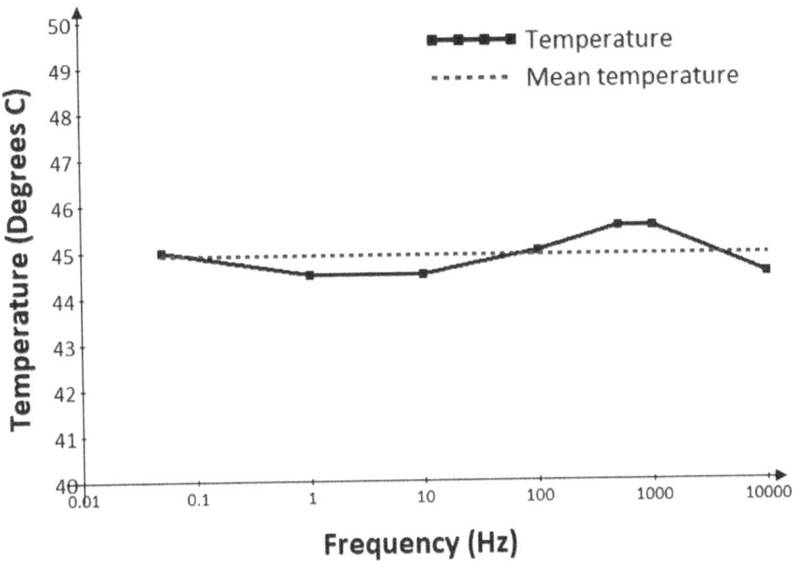

Figure 15.8 Temperature at 50% duty cycle as a function of frequency.

Next, we looked at the temperature as a function of duty cycle at two different frequencies, 100 Hz and 10 KHz. Those results are plotted in Figure 15.9. They reflect a linear relationship, within the probable experimental error. These results are consistent with the conclusions reached above (see Figure 15.1) with the TRM model.

Finally, we set up the test system to record the temperatures for a 0.04-Hz frequency at an 80% duty cycle (after the temperature cycles had had time to stabilize). The results are shown in Figure 15.10. We have also included some data from the TRM PWM analysis (the green curve from Figure 15.5) plotted on the same axes for the same conditions for comparison. Recall that the thermocouple has a 0.5-degree resolution and a 1-second sampling rate. The curve also reflects some wiggles caused by the difficulty in holding the thermocouple steady for several seconds! Also recall that the PWM analysis slightly understates actual temperature because it does not take the thermal coefficient of resistivity into consideration.

15.3.1 Conclusions

It would appear that we can confirm the preliminary conclusions reached above:

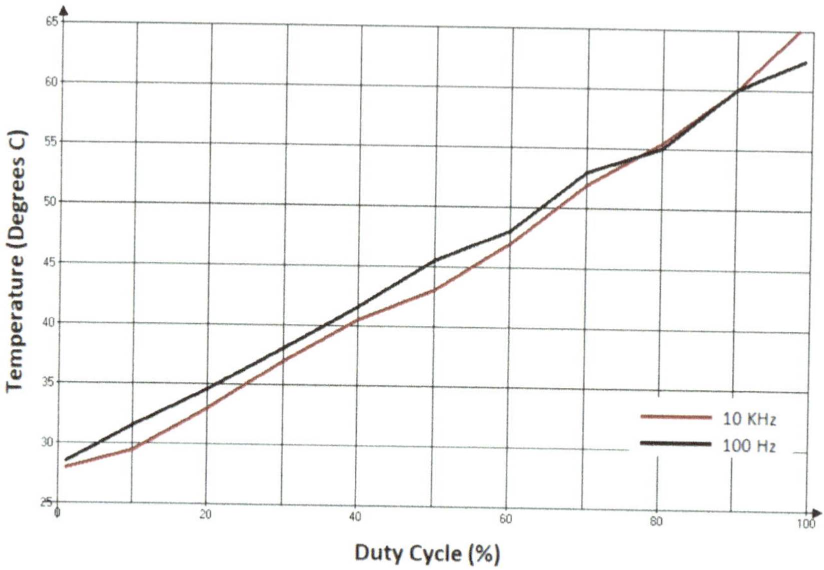

Figure 15.9 Temperature as a function of duty cycle.

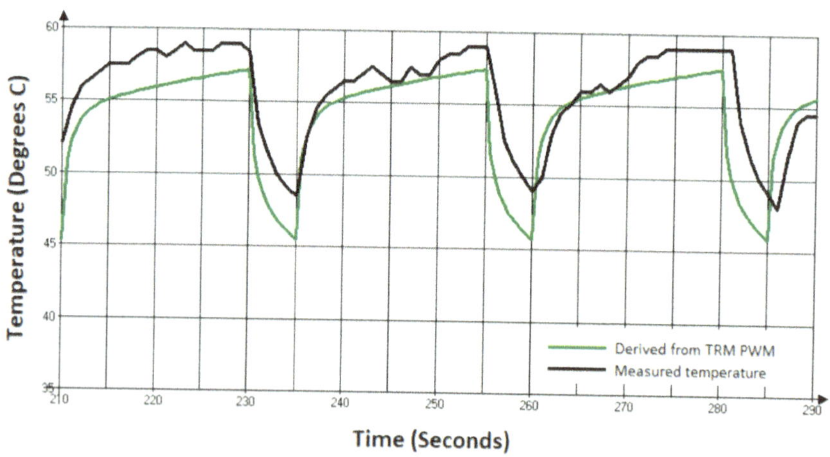

Figure 15.10 Temperature versus time for a 0.04-Hz, 80% duty cycle waveform compared to the model plotted in Figure 15.5.

1. The temperature of a trace is determined by the equivalent RMS value of the current driving it;
2. The temperature of a trace at any frequency and duty cycle seems to be independent of frequency (at least at frequencies below where the skin effect comes into play).

15.4 Analog AC Currents

But a question still remains: What about analog ac currents? What about sine or triangular waveforms? The next section looks at the slightly more complicated topic of analog waveforms to see if the same conclusions apply.

15.4.1 Test Circuit

The first problem is how to apply an analog signal of known, repeatable current level through a trace at a significant enough current to heat the trace. To accomplish this, the circuit shown in Figure 15.11 was created. A waveform generator applied one of three signal types to the base of the Darlington transistor. That resulted in a current through the trace controlled by the current through the collector. That current was supplied by a 10.0-volt constant voltage generator through a 1.05-ohm resistor (with a very low temperature coefficient). The current through the trace is derived as

$$\text{Current} = (10.0 - \text{voltage at collector})/1.05 \quad (15.2)$$

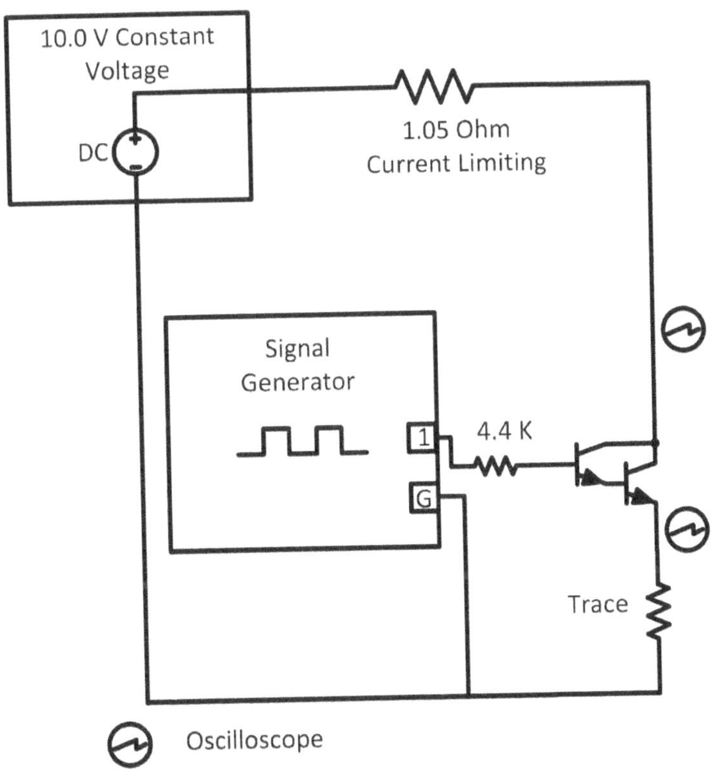

Figure 15.11 Test schematic.

The waveform generator settings were extremely flexible. The output voltage could be set from 0.0 to 10.0 volts. The frequency could be set from 0.01 Hz to over 20 MHz. And the signal DC offset could be set from -10.0 to +10.0 volts DC. Three waveforms were employed during this analysis; sine, triangle (actually sawtooth), and square. The duty cycle for the square and triangular waveforms could be set from 0.0% to 99.9%. The duty cycle for the triangular waveform was set at 99.9% (sawtooth), and the duty cycles for the square waves were set at 50%, 75%, and 99%. Output levels were set to result in large output currents, but not large enough to switch the transistor off nor to bring it into saturation. Samples of the waveform generator output waveforms are shown in Figure 15.12. The generator waveforms exhibit no visible distortion.

15.4.2 RMS Signal Levels

Next, we need to consider how to calculate the rms values of the signals.[2] There are some reference lines drawn on Figure 15.12:

A1 Peak voltage

A0 Minimum voltage

Am Average voltage, that is, Am = (A1-A0)/2)

If we ignore any dc offset, the rms value of a simple square wave is simply its peak voltage, shown as (A1-Am) in Figure 15.12. But if the square

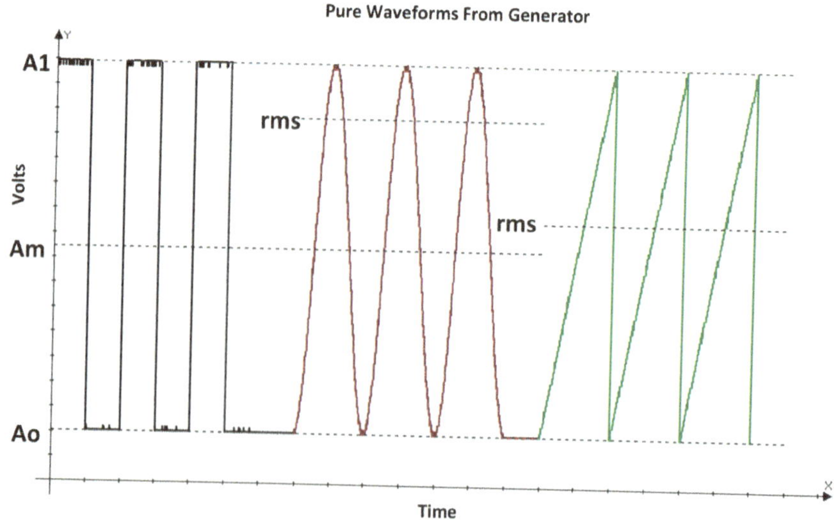

Figure 15.12 Waveform generator outputs.

wave has a duty cycle other than 50%, it looks more like the pulse train described in Section 15.2. The rms value of the pulse train is its peak value multiplied by the square root of the duty cycle, or (A1-Ao)*SQRT(duty cycle). Similarly, if we ignore any dc offset, the rms voltage of a simple sine wave is 0.707*peak, or 0.707*(A1-Am) in Figure 15.12. And, ignoring any dc offset, the rms value of a simple triangular (or sawtooth) waveform is 0.577*peak, or 0.577*(A1-A0) in Figure 15.12.

If there is a dc offset, the rms values are calculated as follows:

Let rms = the rms value of the simple waveform.

Let dc = the dc offset (Am for the simple square wave and the sine wave, Ao for the sawtooth wave and the pulse train.)

Then the rms value of the offset waveform is given by

$$rms_{offset} = \sqrt{rms^2 + dc^2} \tag{15.3}$$

15.4.3 Nonlinearities

But there can be a problem in a practical circuit. The circuit (transistor) is nonlinear over the range of current applied here. For example, Figure 15.13 shows the resulting sawtooth waveform. The degree of nonlinearity is obvious.

The oscilloscope used in this analysis is a digital scope whose output can be saved to a computer file.[3] The scope also has the ability to calculate the

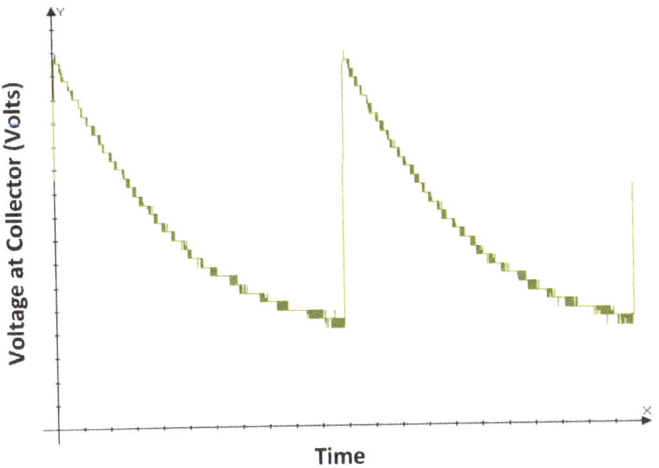

Figure 15.13 Sawtooth voltage waveform vs time at the collector.

"True RMS" voltage (as well as the dc average, and many other measures) of the waveform under evaluation. Therefore, we are freed from having to manually calculate the rms value.

But then, there is another problem. In our circuit; high voltages at the collector correspond to low currents through the trace, and low voltages at the collector correspond with high currents through the trace. The root-mean-square calculation tends to bias the results to the higher values, in this case the opposite of what we want. Fortunately, the scope can output a data file that can be imported into an Excel spreadsheet. The data can be corrected and the rms value calculated by the spreadsheet.[4]

15.4.4 Results

The trace used in this analysis was another trace available to us from our previous fusing investigations. It was 100 mil wide, 0.5 oz thick, and 6 inches long. The resulting summary data is shown in Table 15.3 and graphed in Figure 15.14. These results were obtained at a frequency of 100 Hz. The black line in Figure 15.14 is the current/temperature relationship for the trace with DC current applied. These results are consistent with the prior results (i.e., the effective current for current/temperature analyses of ac waveforms is the rms value of the current).

When applying the signals to the trace, we also varied the frequency from 0.5 to 1,000 Hz. The temperatures remained constant with frequency within the measurement accuracy of the study (as expected). (At frequencies significantly above 5,000 Hz our test setup became unstable.)

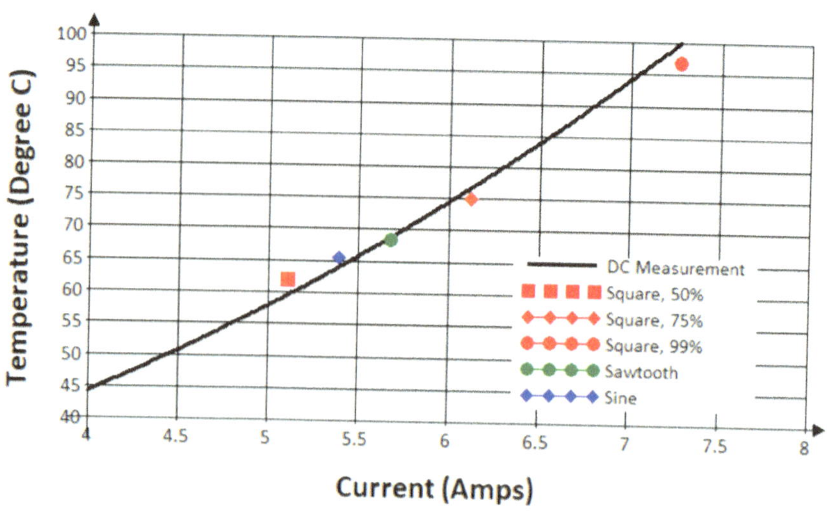

Figure 15.14 Graph of summary results.

Table 15.3
Summary Results of Testing

Waveform	Maximum V	Minimum V	RMS Current (A)	Trace Temperature (°C)
Sine	7.5	2.1	5.39	65.5
Sawtooth	7.15	2.0	5.68	68.5
Sq, 50%	7.0	2.25	5.11	62.0
Sq, 75%	7.68	2.25	6.12	75.0
Sq. 99%	7.9	2.25	7.27	97.0

15.4.5 Conclusion

Again, we conclude that the results of this analysis are consistent with the prior results:

1. The effective current of an analog ac waveform is the rms (root mean square) current;
2. The temperature of an analog ac waveform is independent of frequency (at least over the tested range).

We were not able to test trace current/temperature relationships at very high frequencies because of our equipment limitations. However, theory suggests that the current/temperature relationships are related to the RMS values of the current waveforms. This should be true even at frequencies where skin depth issues come into play. Nothing in our investigations would indicate otherwise.

End Notes

1. At frequencies above about 10 Hz the thermal inertia is such that the temperature does not change much within the cycle.
2. See https://en.wikipedia.org/wiki/Root_mean_square.
3. The oscilloscope used is a PicoScope model 2204A.
4. There are approximately 900 data samples per cycle in our waveforms. The Excel formula for calculating rms is =SQRT(SUMSQ(A1:A10)/COUNTA(A1:A10)) where A1:A10 is the data range of interest.

CHAPTER

16

Contents

Industrial CT (X-Ray) Scanning

16.1 Bottom Line

- Industrial CT scanning can be an important tool in analyzing PCB failures. But it does not offer quite enough resolution (yet) to allow us to replace destructive microsection analysis for measuring trace parameters.

16.2 Background

When trying to compare actual results to simulations, it is important to know the *actual* trace dimensions (not just the nominal design dimensions). The typical way we check the actual dimensions on a PCB is through taking a microsection of a representative section of the trace somewhere on the board. But there are some problems with microsections:

1. First and foremost, they are destructive;

2. Second, they represent a single *point* along a trace, but that point may not be representative of other points along the trace;
3. If we microsection the actual trace we are evaluating, the microsection precludes any follow-up testing we might want to do;
4. If we microsection the same place on the trace, but on another board in the production lot, we are assuming the parameters of all the boards in the lot are identical, which may or may not be true.

So we began asking the question if x-rays would be an alternative way to measure trace dimensions. X-rays would be nondestructive and would allow us to look at any and all points in the three-dimensional volume of the board (including the walls of vias). And it turns out there are services available that can x-ray a board. This section reports on what we found.

16.3 The Promise

Figure 16.1 shows an x-ray of a board where a trace had failed. This kind of analysis allows someone to look into the board at any depth and at any angle (even inside the IC packages) to look for failures. It can look down at

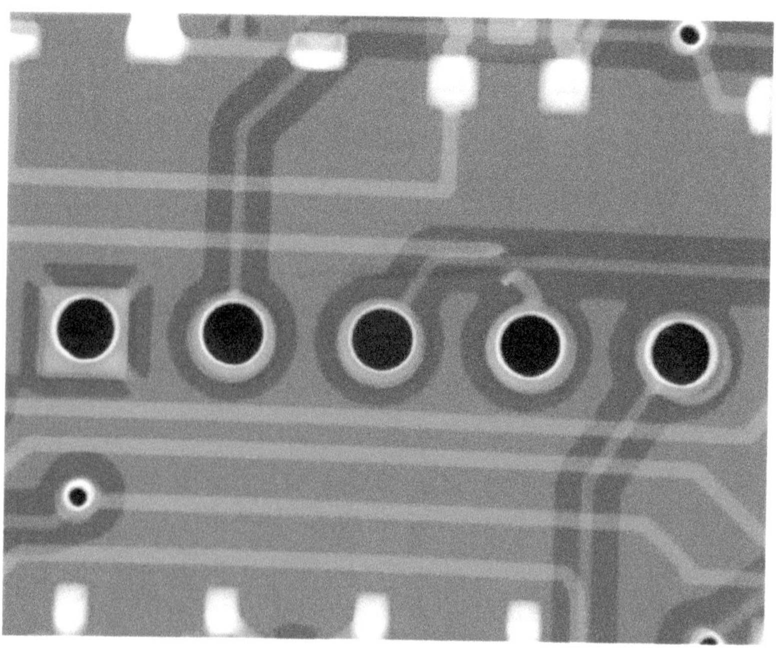

Figure 16.1 X-ray of board showing a failed trace.[1]

traces (as in this view) and check routing parameters, but it can also look at the sides of traces (edge on). It can even look inside vias. So we hoped it could even measure trace thicknesses.

16.4 The Microsectioning Process

Microsectioning starts by punching out a rectangular piece from the board[2] (Figure 16.2) approximately 1 inch long and slightly under 0.5 inch wide. Figure 16.3 illustrates where two such pieces have been removed from one of our test boards. Cut (B) is a typical microsection cutting *across* a trace perpendicular to its direction.[3] But we cut numerous microsections *along* the trace (A) so that we could look for thickness uniformity along the trace. The pieces are then encapsulated in a resin (Figure 16.4) and after about 10 minutes of hardening, they are polished (Figure 16.5). Figure 16.6 illustrates a finished microsection of a via from the board, and Figure 16.7 illustrates a microscopic close-up of the via.

It is possible to scale a view like that shown in Figure 16.7 and measure the dimensions very precisely (better than 0.1 mil or 2.5 μm).

The benefits of microsectioning include the ability to measure dimensions very precisely. But the weakness is that we only get a two-dimensional view of a particular 0.5 in^2 cross section perpendicular to the surface. And again, this process is destructive in its nature.

Figure 16.2 Punching out the microsection.

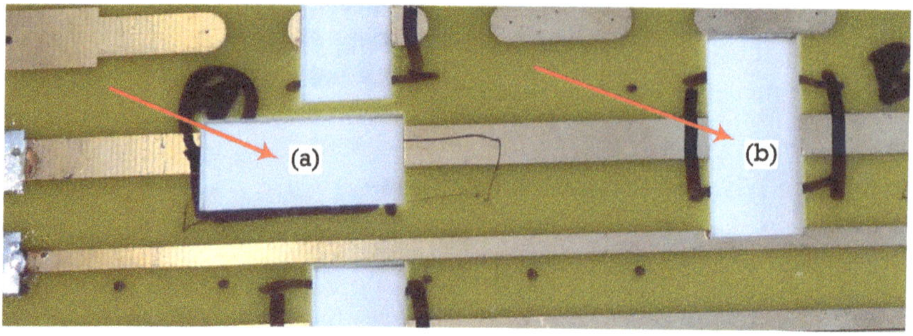

Figure 16.3 Holes in PCB after microsectioning.

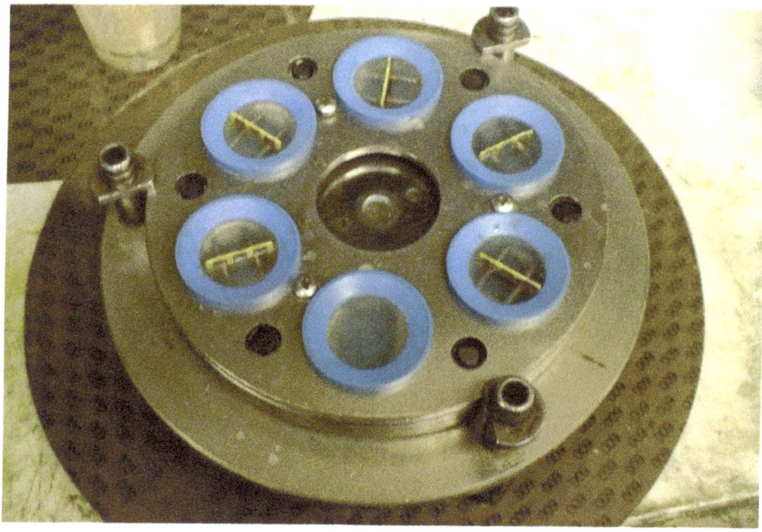

Figure 16.4 Microsections encapsulated in resin.

16.5 Industrial CT Scanning

Industrial CT scanning (also sometimes called computed tomography)[4] is somewhat similar to a medical computed axial tomography (CAT) scan. Referring to Figure 16.8, there is an x-ray source that sends out a beam in a roughly conical shape. The target (think PCB) is placed in the path of the beam. A sensor collects the beam and generates a digital representation of the target. The closer the target is to the source, the larger will be the magnification. But the closer the target is to the source, the smaller is the area of the target that can be analyzed.

Figure 16.5 Polishing the microsection.

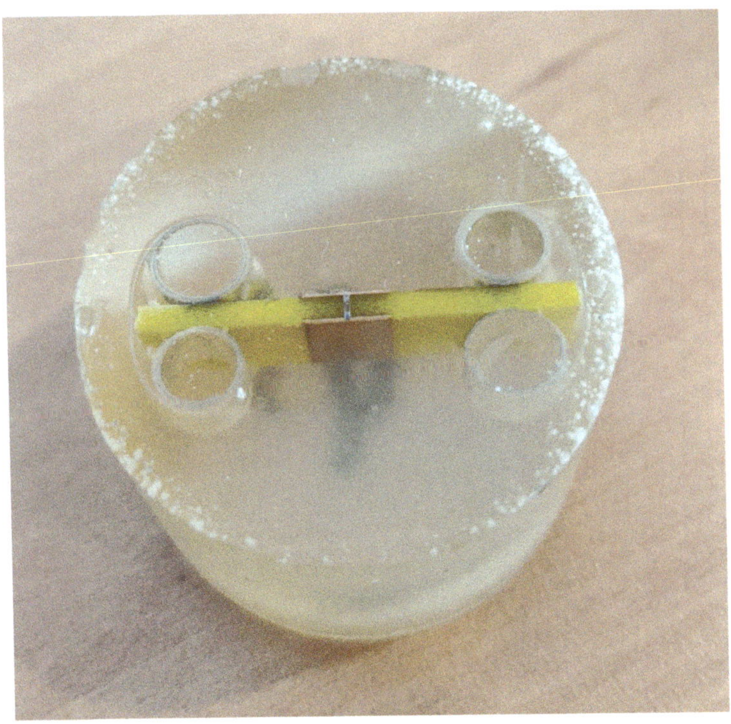

Figure 16.6 Finished via microsection.

Figure 16.7 Enlarged view of a microsection of a via. Dimensions in mils.

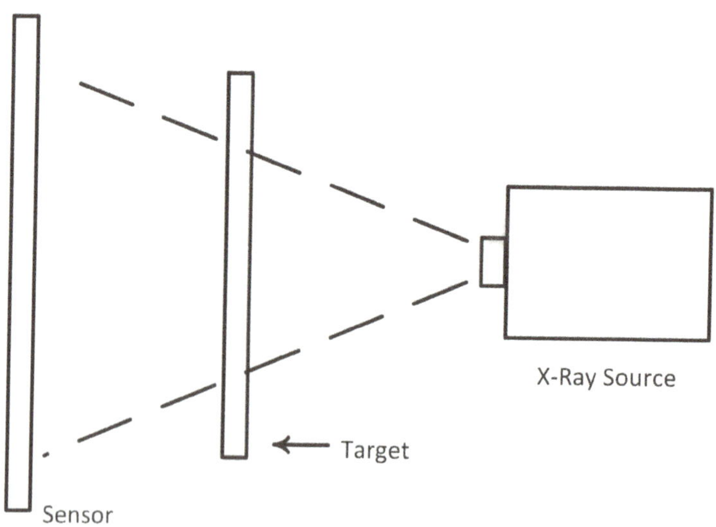

Figure 16.8 X-ray CT scan process.

The target is rotated 360 degrees. Individual x-rays are taken as the target rotates. Roughly 2,200 x-rays are taken during a 360-degree rotation! The magic happens when the software collects all those digital images and combines them into a data file that can be viewed as a three-dimensional image (with a special software viewer). Figure 16.9 illustrates the actual equipment used for generating the x-rays for this project.

16.5.1 Results

Figure 16.10 shows our particular test board. The viewer is positioned at a point with a via exactly like the via shown in Figure 16.7. The bottom right image is that of the board (it was cut down to fit on the target stand. More about that later). The trace under the cross-hair is a short stub with vias at each end. The vias are connected to similar trace stubs on the opposite side of the board.

The other three quadrants provide views in the three primary axes. The upper right quadrant is viewing down onto an x-y plane at a particular point along the z-axis (y is up; x is to the right; z is into the page). In this view, the plane is the middle of the trace slicing through the via. The line points to the via. The upper left quadrant is the view looking back in the y-direction at a point along a plane in the x-axis (y is up; z is to the right; x is into the

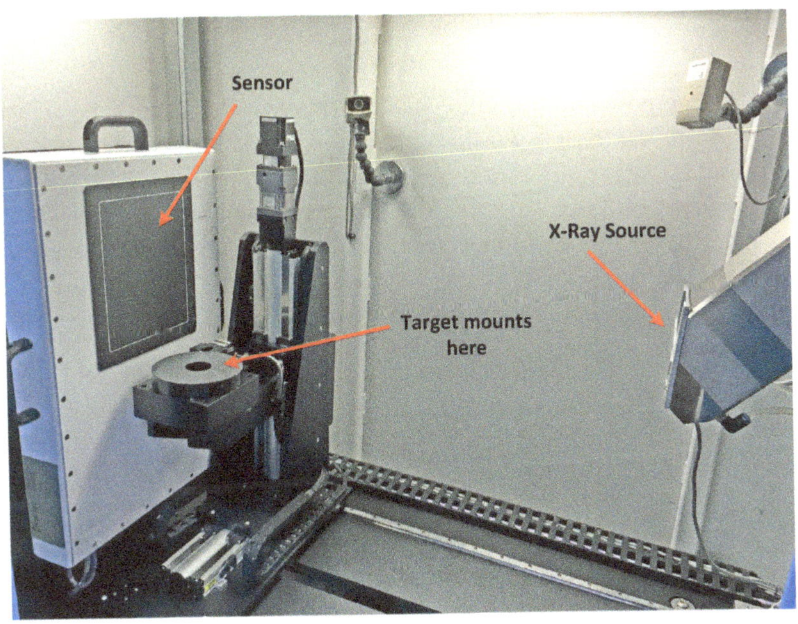

Figure 16.9 X-ray equipment used for this project. (Courtesy Jesse Garant Metrology Center.)

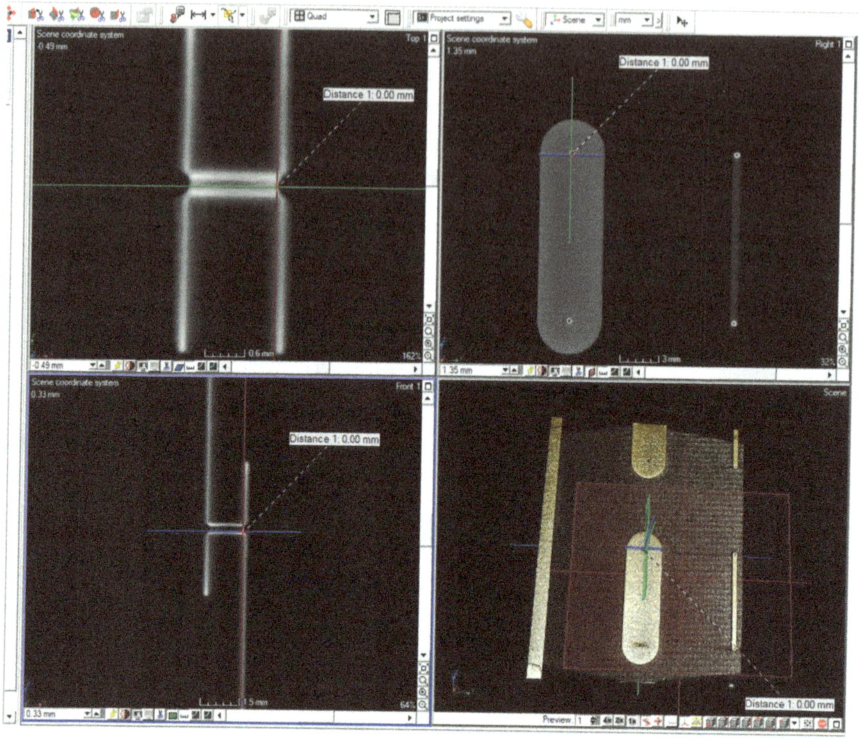

Figure 16.10 X-ray images of our board.

page). The line points to the same spot on the via. The lower left quadrant is looking from left to right on the *x*-axis along plane in the *y*-axis (*x* is up, *z* is to the right). Again, the line is pointing to the same spot on the via.

The image point can be moved to any arbitrary point in three-dimensional space and can be zoomed in or out to reveal any desired area. One of the many tools available in the viewer is the precise measurement between any two points in the board. Two aspects of this ability are particularly intriguing:

1. We can look at any point inside any and all vias;
2. We can examine the entire length of any trace in all three dimensions.

But the problem? The resolution is not quite good enough to take measurements with the required accuracy. The process is excellent for qualitatively analyzing the makeup of the board (including the interior of any components that might be mounted on the board), but not quite suitable for quantitatively measuring the trace parameters. So we can tell if a trace is missing, or if there is a marked difference in the plating thickness on opposite walls of a via. Within reason, we can locate opens and shorts (except possibly hairline cracks). We can evaluate if there are significant nonuniformities in the plating along a trace. We can even evaluate some solder problems. But we can't precisely measure any of those anomalies.

Can the resolution be improved? Perhaps. If the target is moved closer to the x-ray source, then the image is larger and more precise measurements are possible. But a smaller area of the total board is then visible. Here are some possible advances:

1. Increase the power at the source (along with some attendant power management and cost issues);
2. Shorten the wavelength of the source;
3. Increase the sensitivity of the sensor.

All of these are potential solutions, but they are not practical today. We can anticipate they might become practical at some point in the future.

16.6 Comparison of the Processes

As mentioned several times, the microsection process is localized and destructive. The industrial CT scanning (potentially) allows coverage of the entire board and is not destructive. The x-ray process can be used after components have been mounted on the board; the microsection process, probably not.

The resolution for the microsection process is suitable for quantitative analysis; for x-ray, we can really only do qualitative analyses.

The physical board size may be an issue. We can microsection a piece out of virtually any board, regardless of size. There is a limitation, however, on the size of the board that can be placed on the platform for x-ray analysis. Today that size is approximately 5 or 6 inches square, depending on what exactly we want to do. From a practical standpoint there is no thickness limitation for either approach.

The microsection process is very low cost, requires relatively little capital equipment, and can be done by the fabricator. The x-ray process, on the other hand, requires a large capital investment and then requires some time for the access and processing of the image. It is only practical at companies who are specifically equipped to provide the service.

These comparisons are summarized in Table 16.1.

Table 16.1
Comparison of the Two Approaches

Criteria	Microsection	X-Ray
Destructive	Yes; use as representative of other boards in a manufacturing lot	No
Area/volume	2-D small plane, perpendicular to board surface	Anywhere in 3-D space; can look inside vias and components and along traces
Use on assembled boards	Probably not	Yes
Physical size limit	No	Yes
Resolution	Very good (<0.1 mil)	Limited
Turnaround time	Hours	Days
Cost	Cheap	Expensive
Capital cost.	Modest	Very high
In-house potential	Yes	No
Can fabricator offer	Yes	No
Best use	Measure precise trace dimensions at a point	Evaluate trace and via continuity; locate shorted trace

16.7 Conclusion

There are several benefits the industrial CT scanning process offers us in the PCB industry today. And it was very interesting looking into it. It is suitable for troubleshooting the continuity of traces and probably for evaluating the uniformity of the plating of traces and vias. It would almost certainly allow discovery of a precise point of a conductor short circuit inside a board. Unfortunately, the primary benefit we were hoping to achieve (quantitatively measuring the trace) is not practical—yet! But we have hope for the future.

End Notes

1. Figure 16.1 courtesy of Priority Labs, Richardson, Texas.
2. Figures 16.2–16.4 courtesy of Prototron Circuits, now of Tucson, Arizona, who allowed us to photograph their processes.
3. For a picture of such a microsection see Figure 3.4.
4. There are some significant differences between CT scanning and a CAT scan. The output of CT scanning is a data file that allows the user to look anywhere in 3-D space. Heart patients (among others) often undergo a full CT scan. The output of a CAT scan is typically one or more 2-D views of the area of interest in the patient. A simple medical x-ray of a patient produces a single picture. Taking a set of single x-rays of an industrial product is referred to as industrial radiography. Industrial radiography and industrial CT scanning can be done

at much higher power levels than can be done in medical imaging, which means they can scan denser materials with higher resolution than is available in medical imaging.

APPENDIX

A

Contents

Measuring Thermal Conductivity

A.1 Measurement

In order to run accurate simulations, as described in Chapters 6 and 7, one of the necessary parameters is the thermal conductivity coefficient of the PCB dielectric material. Recall from Section 4.4 that PCB materials are anisotropic. That is, they cool better in the in-plane direction than they do in the through-plane direction. We need to know the thermal conductivity coefficients in both directions for reliable modeling.

Data sheets frequently ignore this parameter. If they do include a thermal conductivity coefficient, they frequently only include a single value, and they frequently fail to state which direction the coefficient applies to.

A search on the internet provides anecdotal estimates of PCB material thermal conductivity ranging from 0.3 to 0.8 W/mK (watts per meter degree Kelvin). Wikipedia references at least six different ways to measure thermal conductivity.[1] The people at C-Therm Technologies,[2] in New Brunswick, Canada, were nice enough to support this research effort by measuring the thermal conductivity coefficient of the board material used in Chapter 7.

C-Therm uses a technology known as the Modified Transient Plane Source (MTPS) technique. The process involves a transducer (see Figure A.1) that provides a constant current heat source for the sample. The sample (approximately 1.5 inches square) is placed on top of the sensor. A short burst of heat (for perhaps a second or two) starts to heat the sample.

If the sample is very thermally conductive, the heat will drain away from the sensor through the sample, lowering the temperature at the sensor. If the sample does not conduct heat well (i.e., has a low thermal conductivity), then the sensor will be hotter. Thus the temperature at the sensor is inversely proportional to the thermal conductivity of the sample.

If the sample is placed on the sensor as shown in Figure A.1, the process measures through-plane (*z*-axis) thermal conductivity. To measure in-plane thermal conductivity, multiple samples are grouped together (and compressed for good thermal conductivity) edge-wise on the sensor. Figure A.2 shows the actual through-plane and in-plane measurements of the samples from the materials used in the simulations described in Chapter 7. In Figure A.2 the actual sensor is placed on top of the sample, weighting it down (and thereby making good thermal contact). The base provides a good, controlled heat sink for the measurement process.

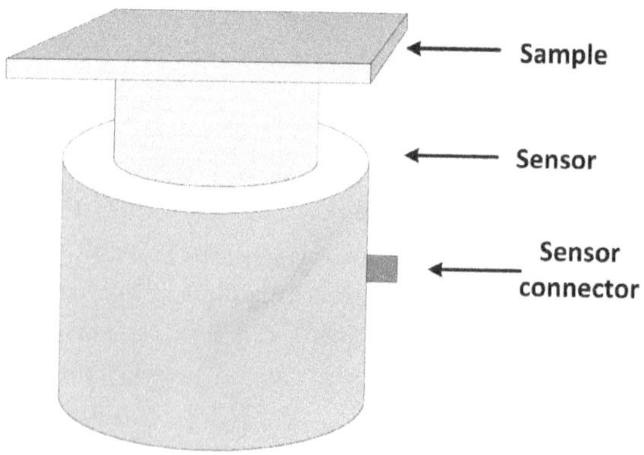

Figure A.1 MTPS transducer.

(a)

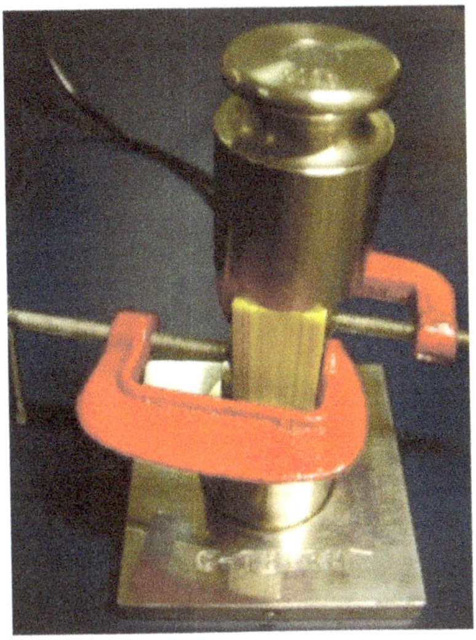

(b)

Figure A.2 Through-plane (a) and in-plane (b) measurements of thermal conductivity for the material employed in the simulations described in Chapter 7.

End Notes

1. https://en.wikipedia.org/wiki/Thermal_conductivity_measurement.
2. http://www.ctherm.com/.

APPENDIX

B

Contents

Measuring Resistivity

B.1 Resistance versus Resistivity

There are numerous anecdotal references to measurements of trace resistivity on the internet with results ranging from about 1.7 to 2.4 μohm-cm. In our opinion most of them suffer from various forms of measurement error. That is why we prepared this short reference on how to measure resistivity.

First, *resistivity* is a parameter associated with an element (such as copper). *Resistance* is a parameter associated with (in our context) a trace. These are different things. The relationship between resistivity and resistance involves the cross-sectional area of the trace and its length and is given by (3.1) (repeated here as (B.1):

$$R = (\rho / A)^* L \tag{B.1}$$

where

R = resistance in ohms

ρ = resistivity in ohms-length

A = cross-sectional area in square units

L = length in units

So, for example, if we change the trace length or cross-sectional area, the resistance will change, but the resistivity of the copper will not.

The resistivity of pure copper is usually given as 1.676 μohm-cm at 20°C. The resistivity of copper alloys can vary, but it is usually given as 1.72 or so. By comparison, silver has the lowest resistivity of any element at 1.6 μohm-cm. So any reported measurement of copper resistivity less than 1.6 μohm-cm (at 20°C) is impossible! There is no element or combination of elements that has this low a resistivity (unless you lower the temperature).

There are anecdotal reports of resistivity measurements of plated copper well above 1.7 μohm-cm (at 20°C) but these results should be suspect because plated copper is *pure* copper, almost by definition. The plating process electro-deposits copper ions (and no other ions) on the cathode. Therefore, plated copper should result in a measured resistivity close to that of pure copper.

B.2 How to Measure PCB Trace Resistivity

The practical way to measure trace resistivity is to start by measuring trace resistance. Then use (B.1) to calculate the resistivity. There are two practical ways to measure trace resistance. One is to apply a known current to the trace and measure the voltage across the trace (or vice versa). The other is to use a very sensitive ohmmeter. We will show that these are actually the same thing!

If you want to measure the resistivity (in fact if you want to measure anything), it is usually beneficial to use the largest parameters practical. By that I mean you want the largest voltage and current practical. You want the largest cross-sectional area practical. And you want the longest length practical. The reason is that it is generally true that larger parameters can be measured more precisely than smaller parameters can. You can usually measure inches to a finer number of decimal places than you can measure mils. You can usually measure hundreds of mV more precisely than you can measure tens of mV.

And then, where and how you place your probes is very important in measuring resistance. Figures B.1 and B.2 show two different probe attachment methods for applying a current to a trace and measuring the voltage across the trace.

The reason Figure B.1 is wrong is because the measurement of the voltage across the trace includes the section of the wire lead carrying the current between the lower connection point and the trace. In effect, you are

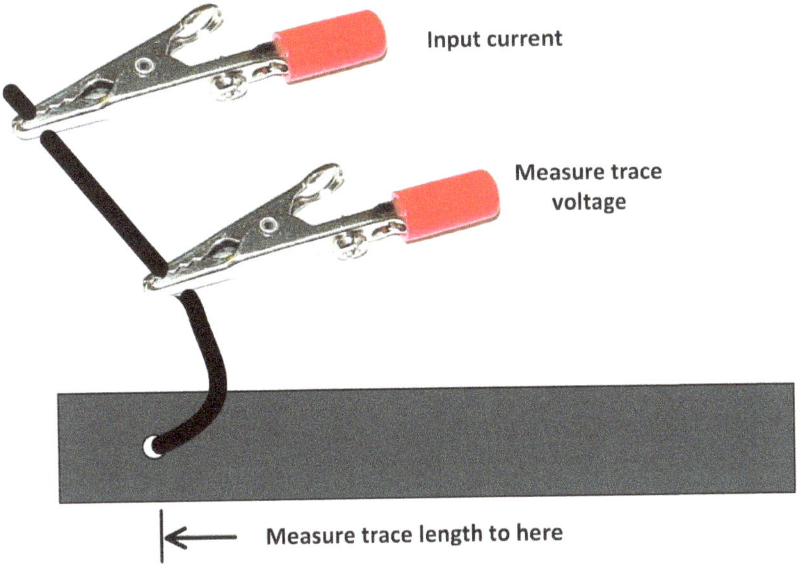

Figure B.1 Incorrect method for measuring trace resistance.

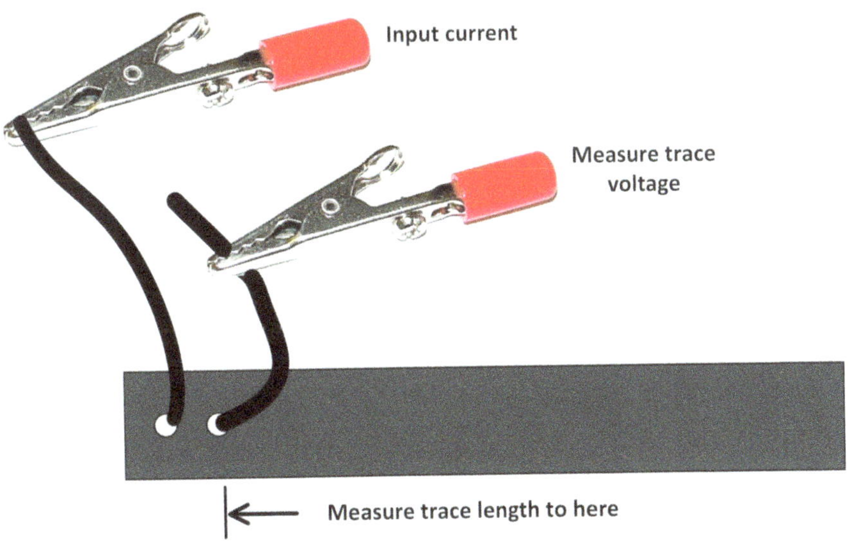

Figure B.2 Correct method for measuring trace resistance.

measuring the voltage across the trace, part of the wiring, and at the contact resistance between the lead and the trace. If you are using a probe of some type, the situation is worse; probes have a contact resistance that is often a function of the applied pressure. Perhaps you have seen this when you have measured resistance using an ohmmeter.

The method shown in Figure B.2 avoids this problem. In fact, the method shown in Figure B.2 is very similar to the method the IPC uses in their thermal testing (reference Figure 4.5, the IPC standard test trace).

B.3 Problem with Ohmmeter Measurement

Figures B.3 and B.4 illustrate two common forms of ohmmeter configurations, series and shunt.[1] In the series design, R1 and the adjustable resistor are selected so that when the resistor under test (Rx) is shorted the meter reads full scale. This is defined as the zero-resistance point on the scale. When Rx is infinite (i.e., open), no current flows. When Rx equals R1 plus the adjustable resistance, the meter reads half scale.

In the shunt-type configuration, when Rx is shorted, no current flows through the meter. When Rx is open, then R1 is adjusted so the meter reads full scale. When Rx equals R1, the meter reads one-half scale.

The problem with typical ohmmeters is that the current through Rx is usually quite small (meter movements are typically 50 to 200 μA full scale), violating the principle we postulated above that the maximum practical parameters (in this case current) should be used.

There are four-terminal ohmmeters available. Some people believe that using a four-terminal ohmmeter solves any ohmmeter problems with these types of measurements. The difference between two-terminal and four-terminal meters is illustrated in Figure B.5. But note that the difference between the measurement techniques shown in Figure B.5 is *exactly* the same as the difference in the measurement techniques shown in Figures B.1 and B.2! Figure B.5 looks just like the technique of applying a known current through the unknown resistance and measuring the voltage across it. But the problem of small signal levels still remains.[2]

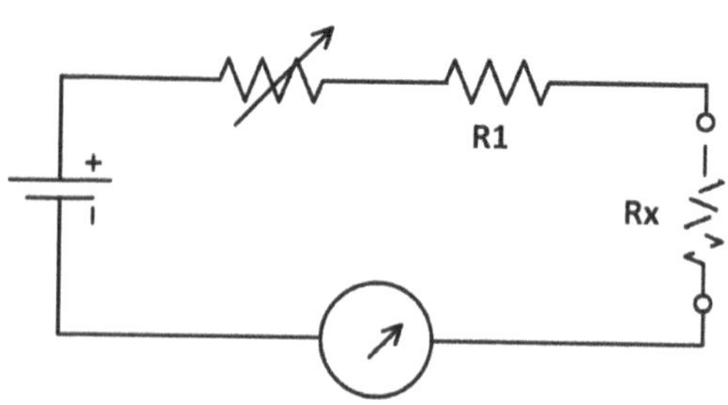

Figure B.3 Representative series-type ohmmeter design.

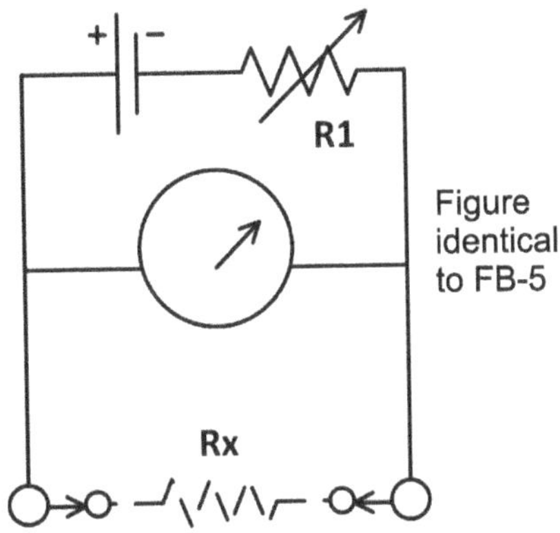

Figure B.4 Representative shunt-type ohmmeter design.

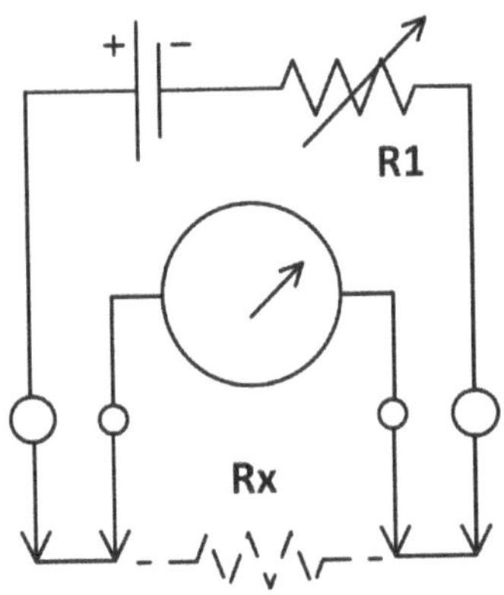

Figure B.5 Typical four-wire ohmmeter.

B.4 Sources of Measurement Error

It is one thing to measure trace resistance. But then we need to calculate resistivity from there using (B.1). In order to do that we need to know the width of the trace, the length of the trace, and the thickness of the trace. *We cannot take the nominal design dimensions for these values.* The three dimensions are discussed below.

B.4.1 Trace Width

Trace width is probably the easiest dimension to deal with. We actually can use the design dimension here. The tolerance on column width should be within 0.1 mil, which results in a measurement tolerance of about 1.0% in even the worst of cases.

B.4.2 Trace Length

Length is a little trickier. Figure B.2 shows where the length should be measured from, and if the trace looks like that shown in the figure, there should not be much of an issue. But if there is a pad at the end of the trace (see Figure B.6), there can be a problem.

The pad can be either circular or rectangular. If the contact point (circle) is at the forward edge of the pad, the measurement will probably work out. We have found experimentally that if the pad is *very* much wider than the trace, then the pad resistance is probably so low that there is no issue, either. But if there is a pad (similar to that shown) and the pad width is not very much wider than the trace, then there will be current densities that will be very difficult to adjust for. The resistance of the pad will impact the measurement.

B.4.3 Trace Thickness

But by far the most difficult issue to deal with is trace thickness. If there is copper plating on the trace, trace thickness can vary by as much as 50% at various places around the board. A 50% on trace thickness results in a 50% tolerance on the cross-sectional area calculation, which will result on a 50% tolerance on the resistivity calculation. Even copper foil typically has a 10% tolerance.

One way to deal with this is to microsection the board where the measurements are being made. But microsections are only valid at the point where they are taken. We can only hope they are representative of other places along the trace. We will show in the next section that trace thickness even varies across the trace at any given point!

B.4.4 Roughness

Figure B.7 illustrates the roughness between the copper foil and laminate for a 2.0-oz trace. It is our experience, through measurement and simulations, that roughness does not have a significant impact on DC trace resistance.[3]

Figure B.6 Trace and pad.

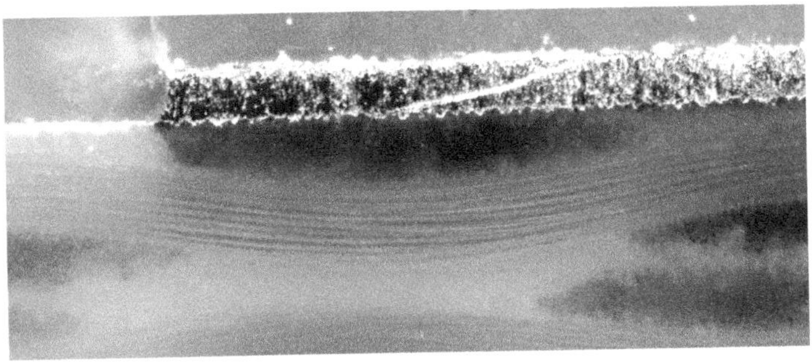

Figure B.7 Roughness between a 2.0-oz foil and the board laminate.

B.5 An Experimental Study

Prototron Circuits was generous enough to donate a board (and microsectioning services) for evaluating trace resistivity. The design criteria incorporated the idea that all dimensions should be as large as practical. The design criteria were as follows:

- Width: Traces were 200 mil wide
- Length: Traces were designed to look like Figure B.2
  - Trace length was 5.0 inches
- Thickness: Three thicknesses were employed:
  - Copper foil (2.0 oz, nominally 2.3 mils thick)
  - Copper foil + 1.0-oz plated copper (total nominally 3.9 mils thick)
  - Copper foil + 1.0-oz plated copper, + 1.0-oz additional plated copper (total nominally 5.2 mils thick).

Traces were grouped in sets of three (the three thicknesses) and repeated around the board. Five sets of traces were analyzed, four on the top layer and one on the bottom layer.

Each set of traces was microsectioned. Figure B.8 is a microscopic view of the end of one of the three-layer, 4.0-oz traces. The lines between the three trace layers are visible in this view. It is also evident that the trace is thicker at the edge than it is a few mils in from the edge, illustrating the fact that the trace thickness can vary even from edge to edge. Figure B.9 is a microscopic view of a two-layer, 3.0-oz trace. The difference in thickness between the edge of the trace and near the midpoint is also evident in that trace. In general, foil traces were pretty flat. Traces with copper plating ap-

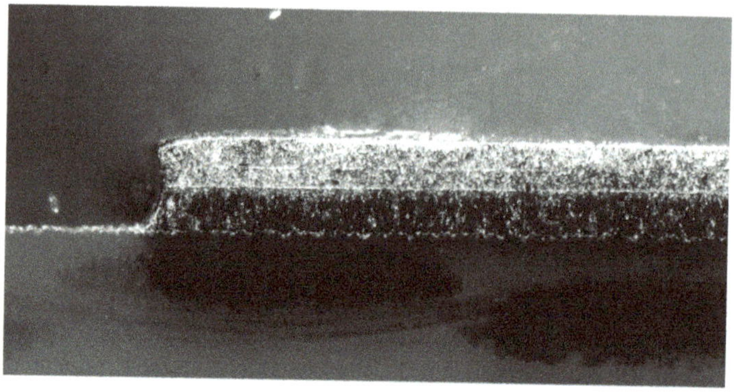

Figure B.8 Three-layer trace.

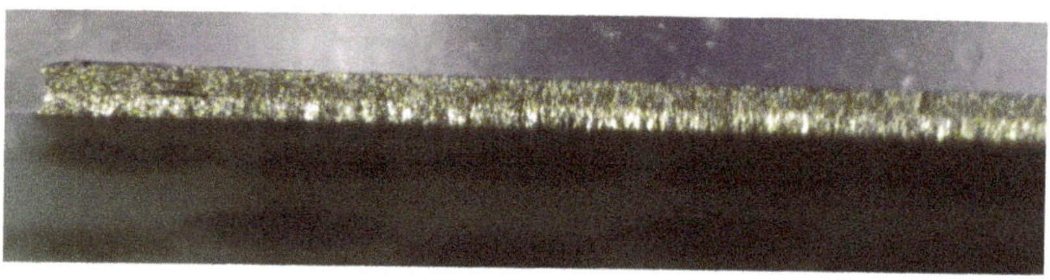

Figure B.9 Two-layer trace.

plied often had some variation in thickness from edge to edge, as much as 10% in some cases.

The thickness of each microsectioned trace was measured at six places across the width. The extremes of the measurements (across all traces) are shown in Table B.1. The measurement tolerance is approximately 0.2 to 0.3 mils. As can be seen there is a very wide variation in thickness for traces that are nominally the same.

A current of 2.1 amps was applied to each trace and the voltage across the trace was measured. This level of current was sufficiently high to allow reasonably precise voltage measurements while not high enough to heat the traces. That is, the traces remained at the ambient room temperature. The resistance was calculated from current divided by voltage (Ohm's law), and the resistivity calculated (based on the nominal width and length, and the average measured thickness of each trace). The resistivities were averaged for each thickness. The resistivities were then adjusted to 20°C (the room temperature was 24.5°C when the data was taken) using (3.6). The actual resistivities were compared to the expected resistivities based on the hypothesis that the resistivity of the plated copper was that of pure copper, 1.7 μohm-cm. Table B.2 summarizes the results.

Table B.1
Thickness Variation of Traces (in Mils)

	2.0 oz	3.0 oz	4.0 oz
Maximum	2.8	4.3	5.7
Minimum	1.9	3.0	4.3
Overall Average	2.4	3.6	4.7

Table B.2
Calculated Resistivities, μohm-cm

	2.0 oz	3.0 oz	4.0 oz
Average Resistivity	1.806	1.786	1.766
Adjusted to 20°C	1.788	1.758	1.739
Expected Resistivity	1.788*	1.752	1.739

*Taken as the expected value for the 2.0-oz foil.

B.5.1 What Is Expected Resistivity?

Think of the three layers of copper as parallel resistors. The first layer is the copper foil. It has an adjusted resistivity of 1.788 μohm-cm. Let's assume the second layer is pure copper. It would have a resistivity of 1.7 μohm-cm. In terms of parallel resistors, we have one resistor (2.0 oz thick) of one resistivity, and a second parallel resistor of resistivity 1.7 μohm-cm. They would combine to have an expected resistivity as shown in the table. We compare that to the measured resistivity (adjusted to 20°C) and would expect the values to be the same if our assumed value for the resistivity of plated copper is correct. As can be seen the values agree very closely.

B.6 Summary

There is no evidence from this empirical investigation that the resistivity of PCB traces is anything other than what would be expected. In particular, the plated copper appears to have a resistivity of 1.7 μohm-cm, that of pure copper.

End Notes

1. See https://www.edgefx.in/ohmmeter-circuit-and-types-of-ohmmeters/ for a description of series and shunt ohmmeters.
2. There are special purpose four-terminal meters available that can supply relatively large currents (2.0 or 4.0 amps) to the resistance under test.

3. It is known, of course, that roughness has an impact on high-speed traces and designs. But our context here is DC.

APPENDIX

C

IPC Internal and Vacuum Curves Fitted with Equations

The following curves comprise:

1. Solid lines: Curves derived from the IPC data.
2. Dotted lines: Curves plotted from the equations developed in Chapter 5 and summarized in Appendix D.
3. Red points plotted on/near a curve: Simulated point for the same trace.

The following legends apply to all sets of curves

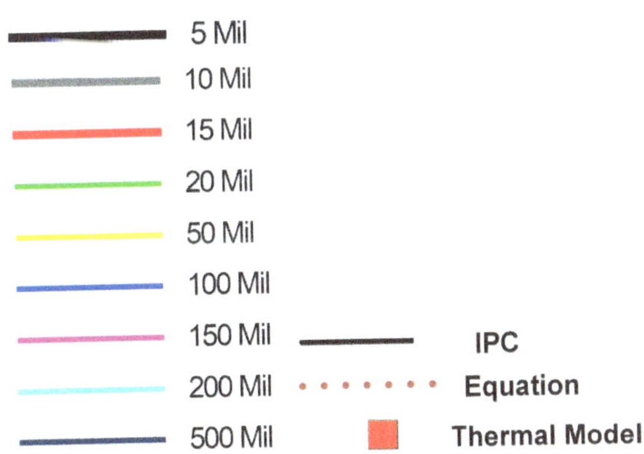

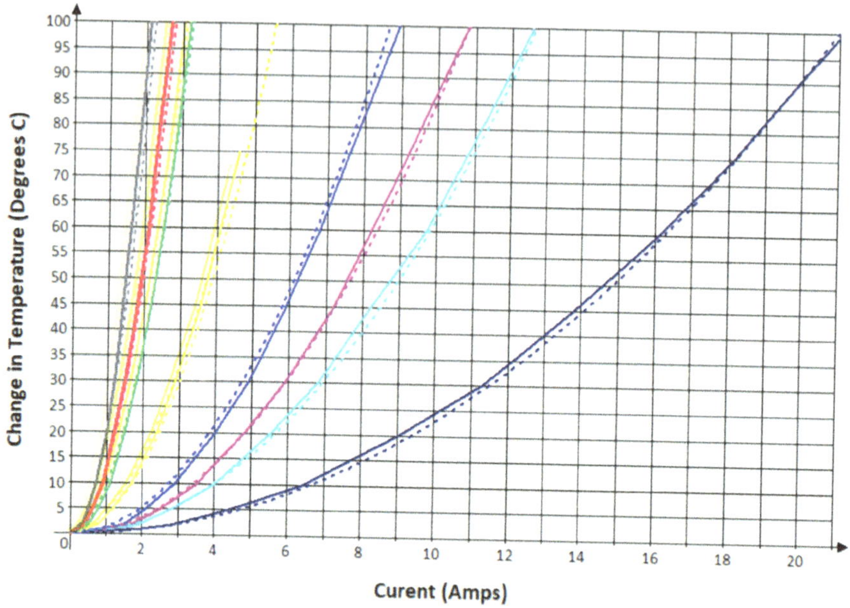

Figure C.1 0.5-oz internal traces.

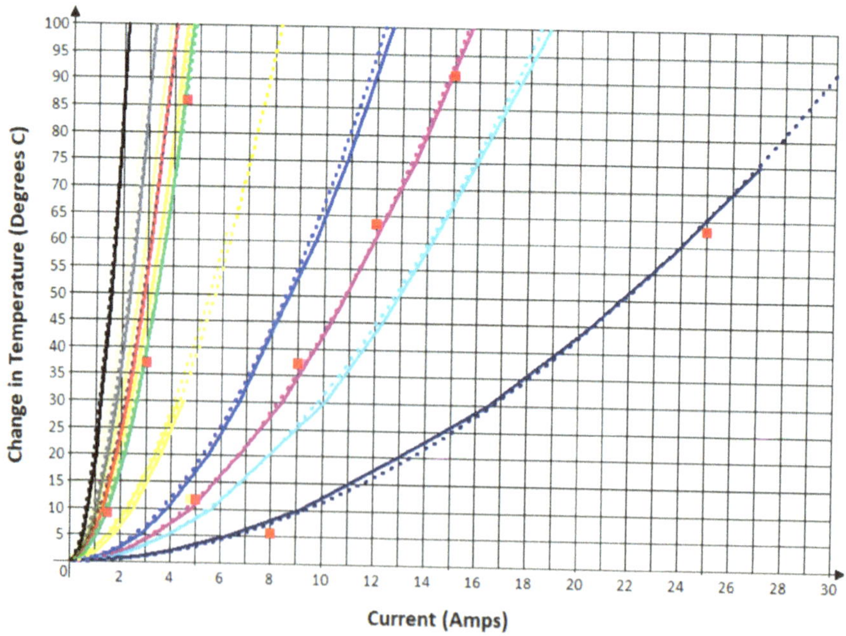

Figure C.2 1.0-oz internal traces.

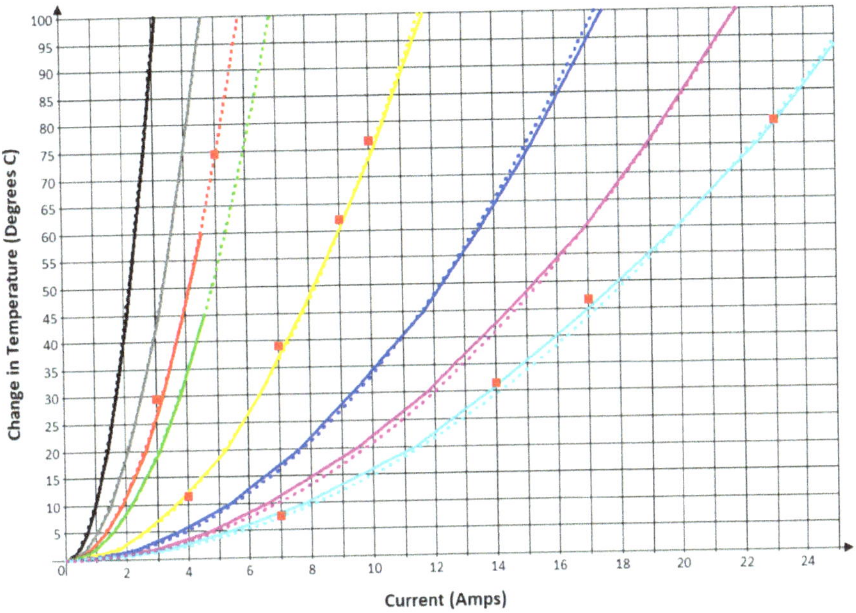

Figure C.3 2.0-oz internal traces.

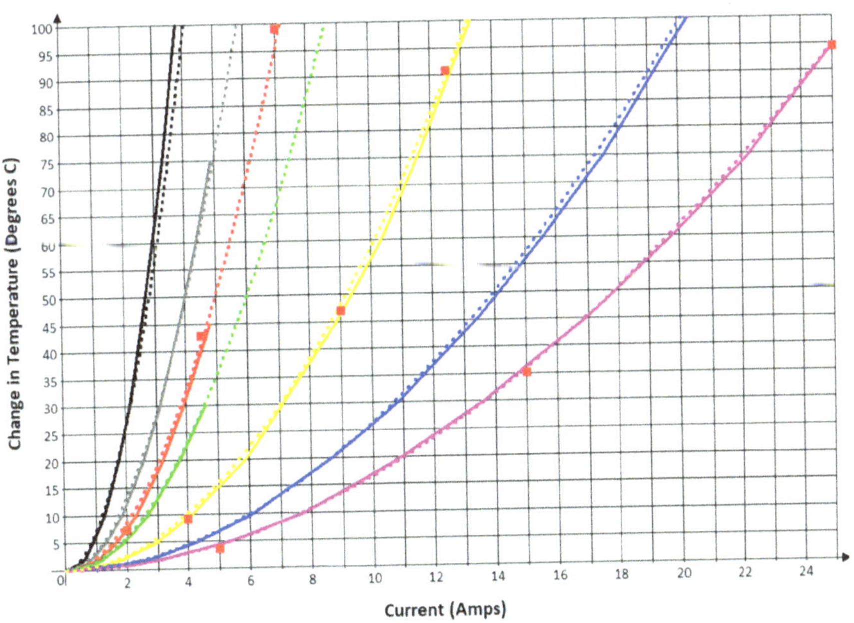

Figure C.4 3.0-oz internal traces.

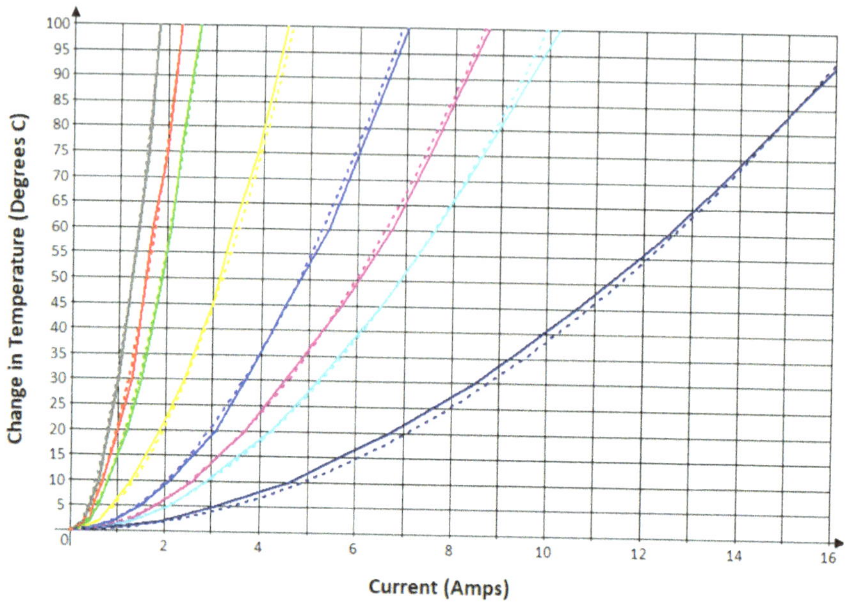

Figure C.5 0.5-oz vacuum traces.

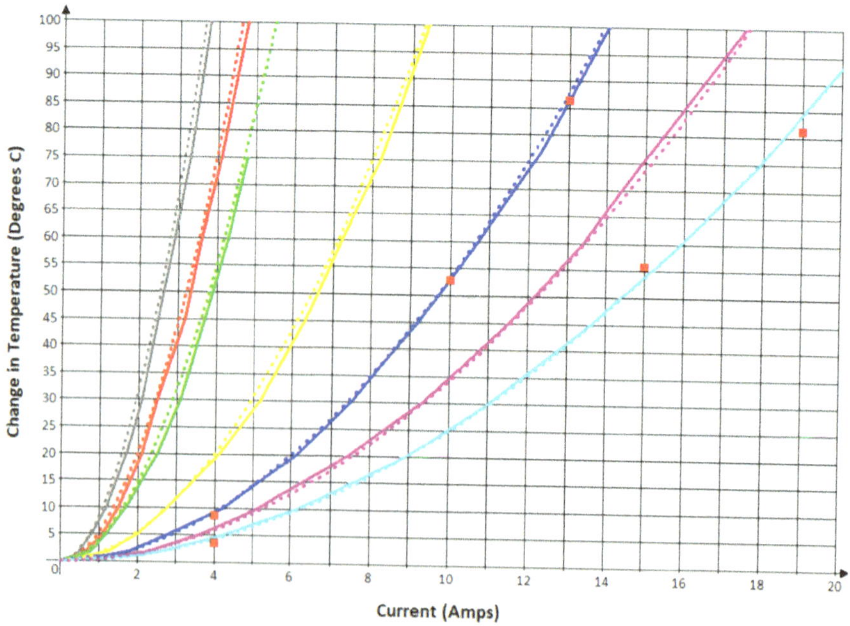

Figure C.6 2.0-oz vacuum traces.

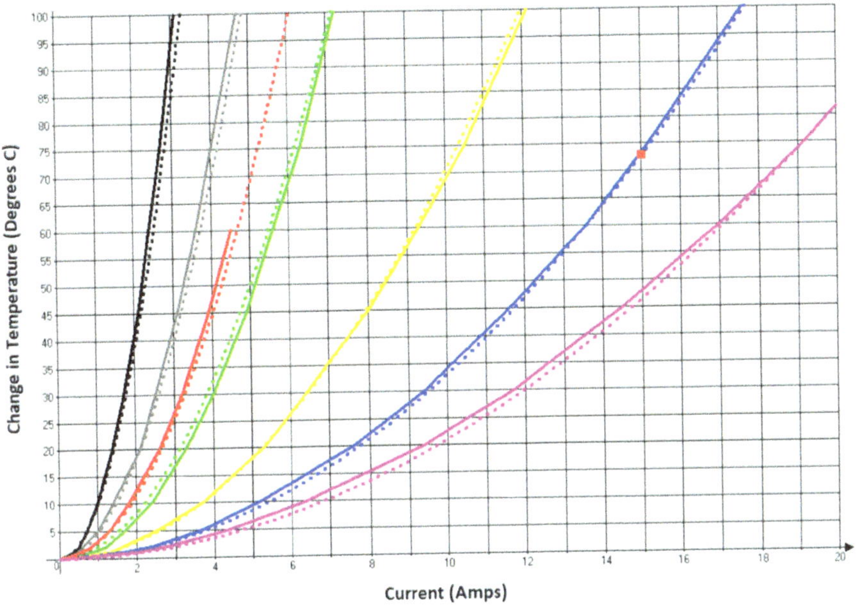

Figure C.7 3.0-oz vacuum traces.

APPENDIX

D

Detailed Set of Equations for the Curves

In the few cases where there are differences by width within a trace thickness, the differences are small and probably reflect errors, uncertainties, and variations as a result of various graphical drawings and manipulations, and material variations that are very difficult so see and determine.

Equations are of the form:

$$\Delta T = \text{constant} * W^{(a1)} * Th^{(a2)} * C^{(a3)}$$

Trace Format	Formula Coefficients				Conditions
	Const.	a1	a2	a3	
External Traces					
All thicknesses	215.3	−1.15	−1.00	2.0	For all widths
Internal Traces					
0.5 oz	110	−1.10	−1.52	2.0	For 100 mil and wider
	125	−1.10	−1.52	2.0	50 mil
	130	−1.10	−1.52	2.0	For 20 mil and smaller
1 oz	200	−1.10	−1.52	1.9	For all
2 oz	300	−1.15	−1.52	2.0	For all
3 oz	429	−1.10	−1.52	1.9	For 50 mil and greater
	368	−1.10	−1.52	1.9	For < 50 mil
Vacuum Traces					
0.5 oz	210	−1.10	−1.52	1.9	For 100 mil and smaller
	215	−1.10	−1.52	1.9	For 150 mil
	225	−1.10	−1.52	1.9	For 200 mil
	235	−1.10	−1.52	1.9	For 500 mil
2.0 Oz.	480	−1.10	−1.52	1.9	For all widths
3.0 Oz.	460	−1.10	−1.52	1.95	For all widths

APPENDIX

E

Current/Temperature Curves, 0.25 to 5.0 oz

The following charts are based primarily on the work done in Chapters 5 and 6. All thermal modeling was done with a model based on the following parameters (English unit conversions are approximate):

- Dimensions: 200 × 110 × 1.6 mm (7.9 × 4.3 in × 63 mils)
- Board material: polyimide
  - $Tcon_{xy}$: .58
  - $Tcon_z$: .58
- Copper ρ: 1.75×10^{-8} Ω-m
- Copper α: .00394
- Trace length: 150 mm (5.9 in)
- Trace width: varies, but constant for each test
- Trace thickness: varies, but constant for each test
- HTC: 10–14, depending on external trace temperature

Each page has a graph for a particular trace thickness (in ounces). There are two sets of graphs, one for a wide current range and another for a close-up current range. A legend is provided for trace thickness for each full set of graphs.

The results obtained with these charts are only approximate. The final design may have characteristics that differ from these results. The user uses these charts at his or her own risk.

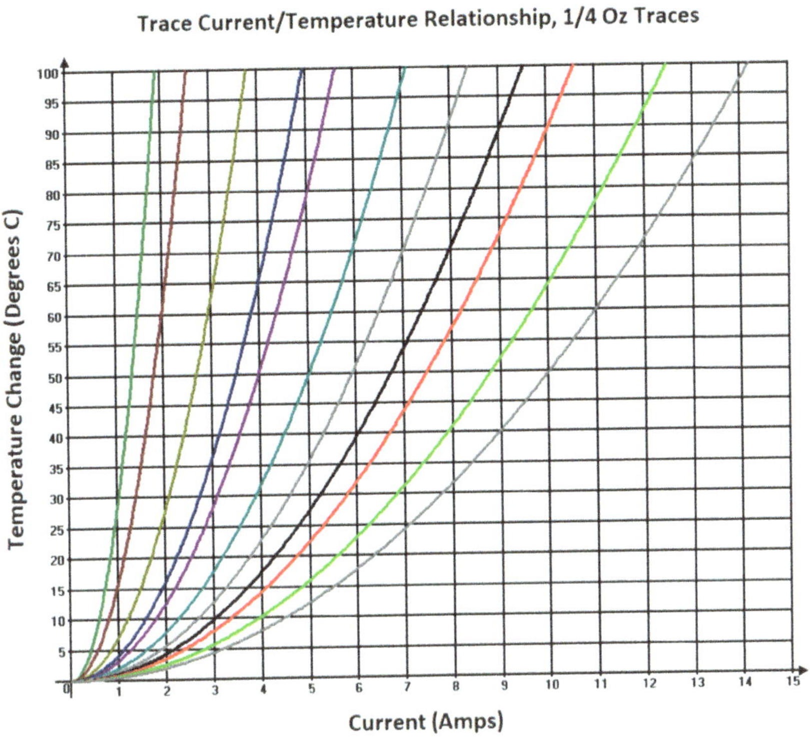

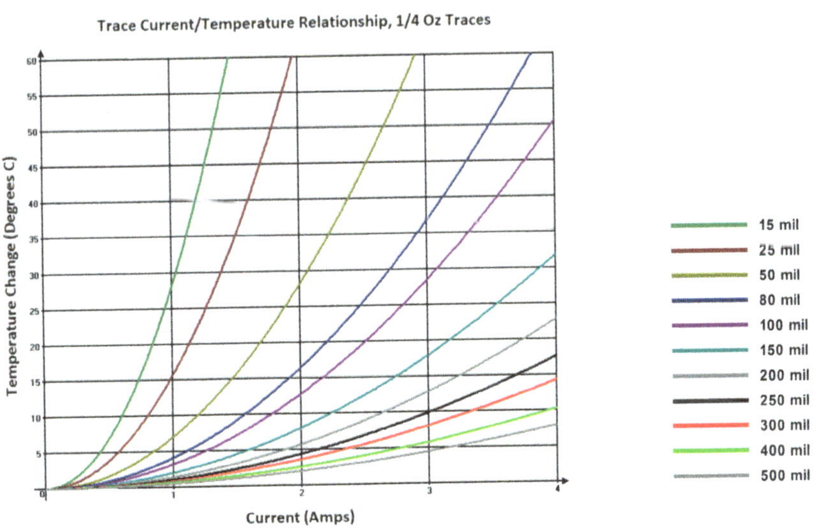

Figure E.1 0.25 Oz

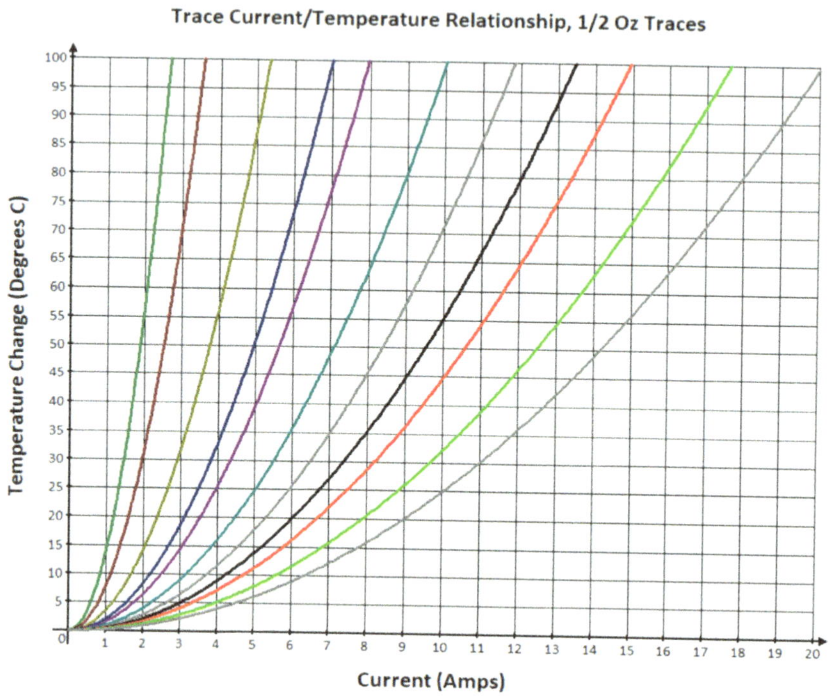

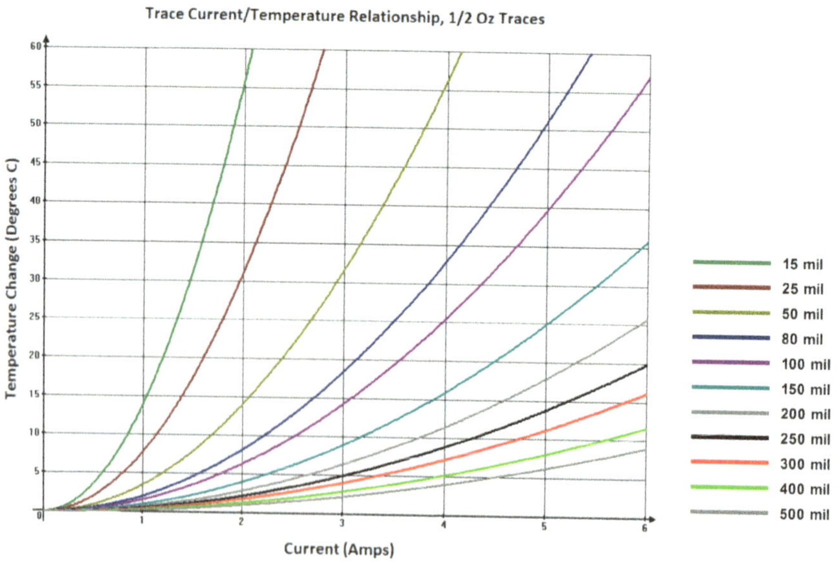

Figure E.2 0.5 Oz

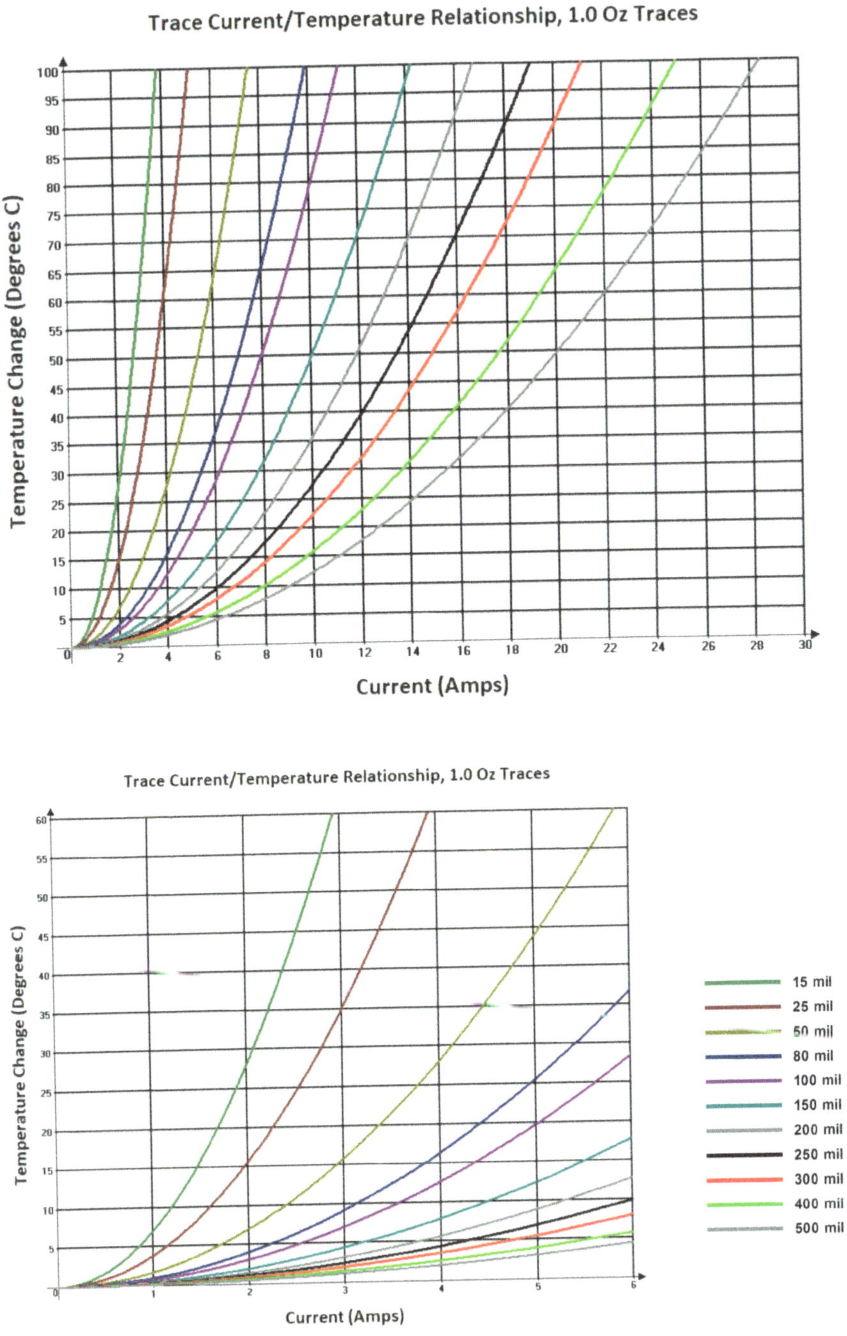

Figure E.3 1.0 Oz

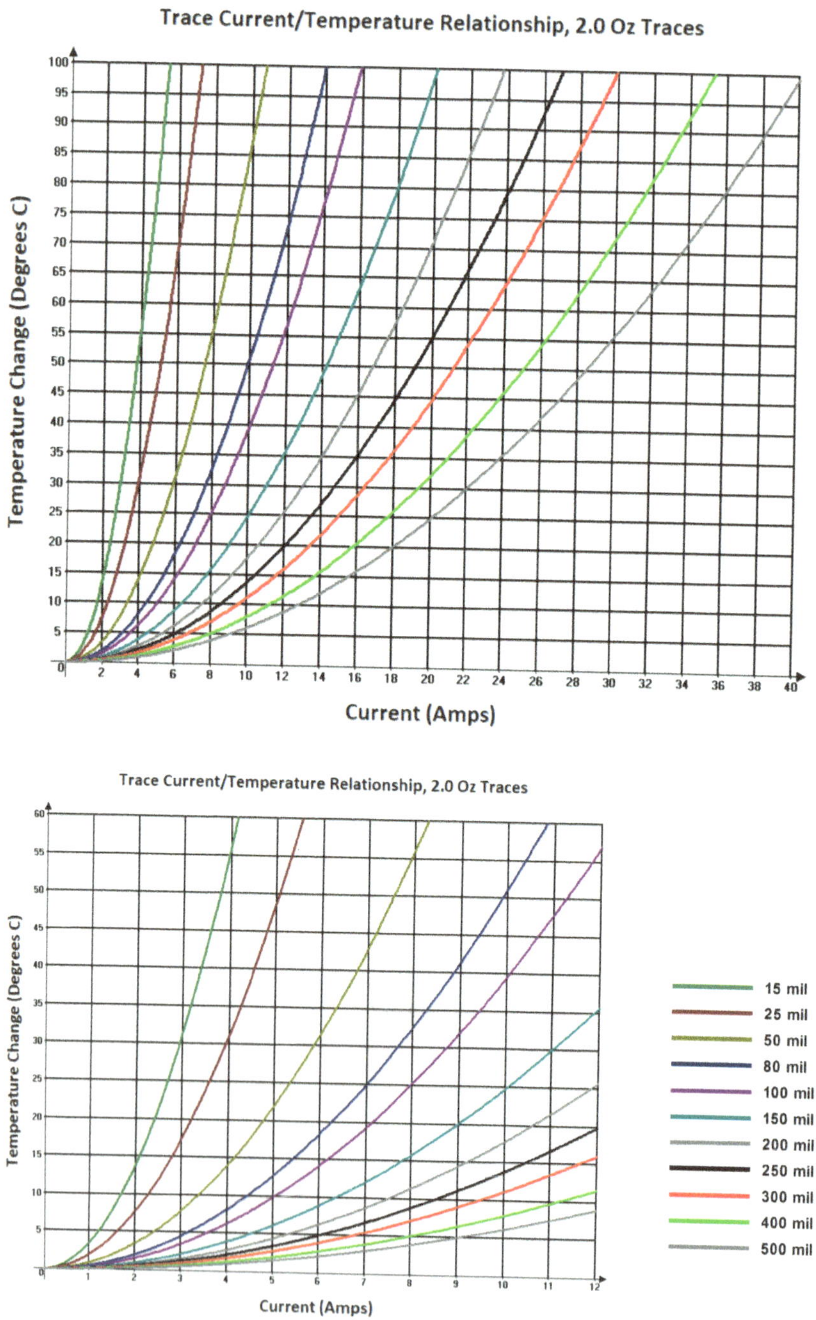

Figure E.4 2.0 Oz

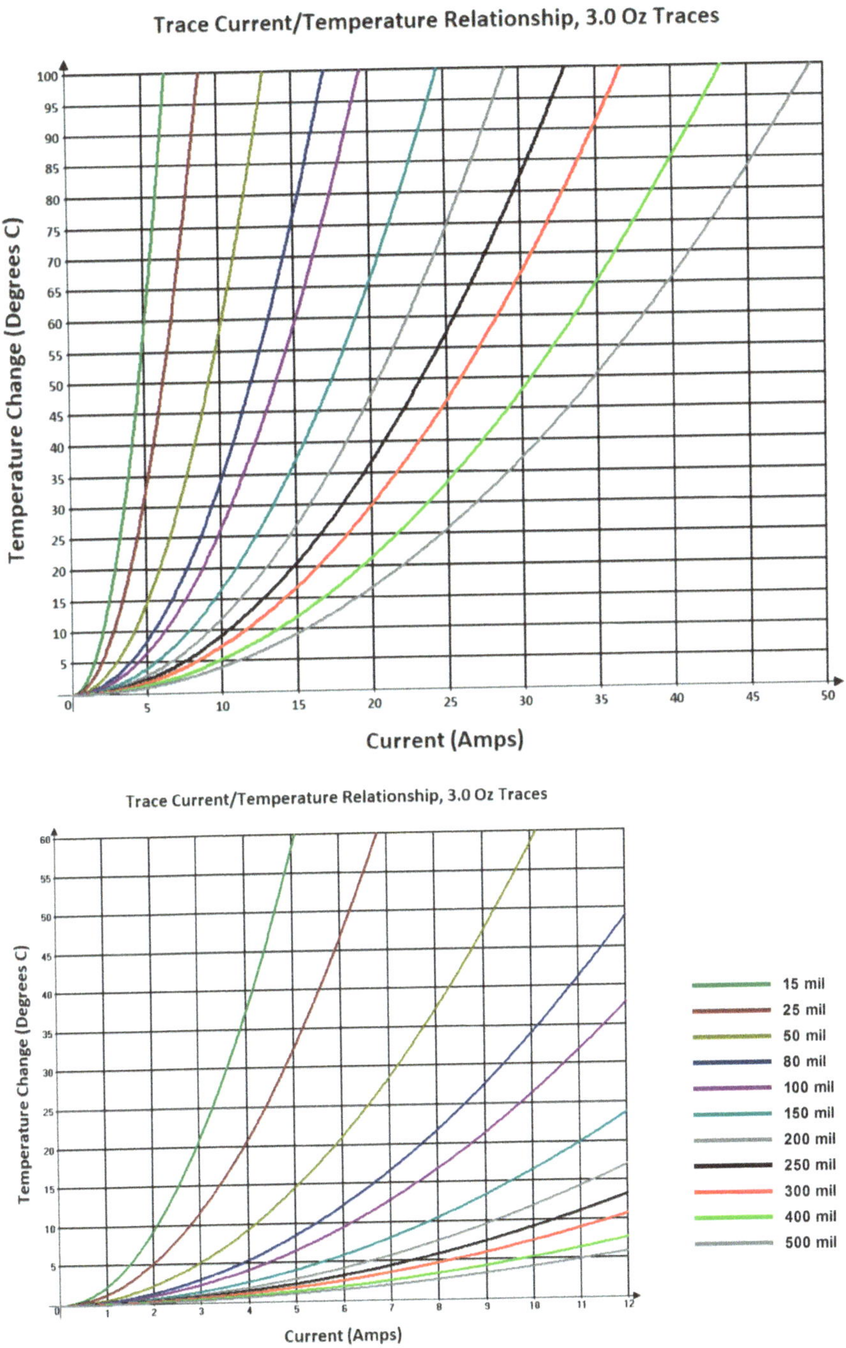

Figure E.5 3.0 Oz

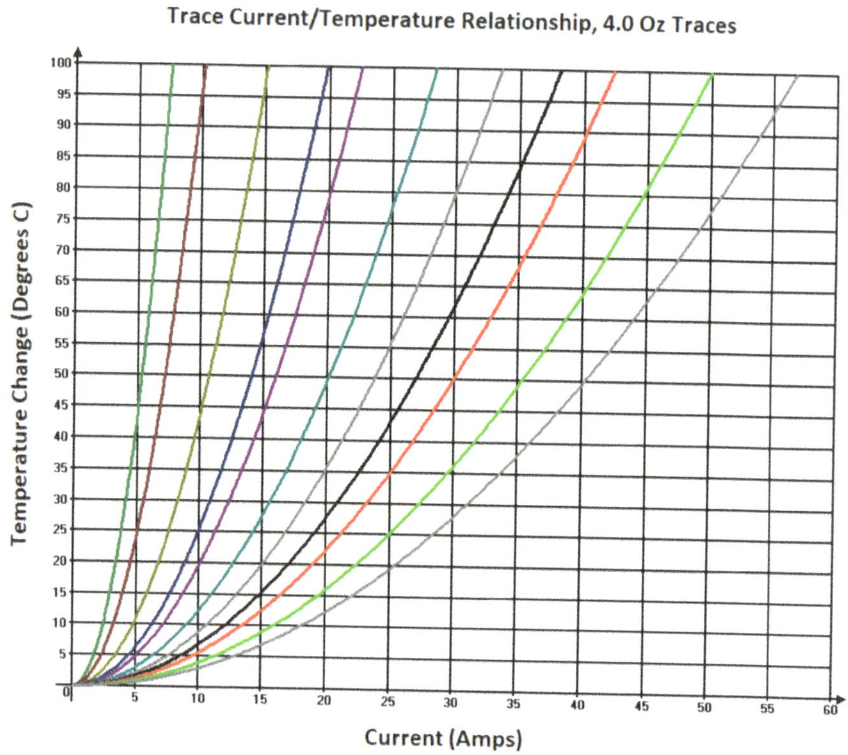

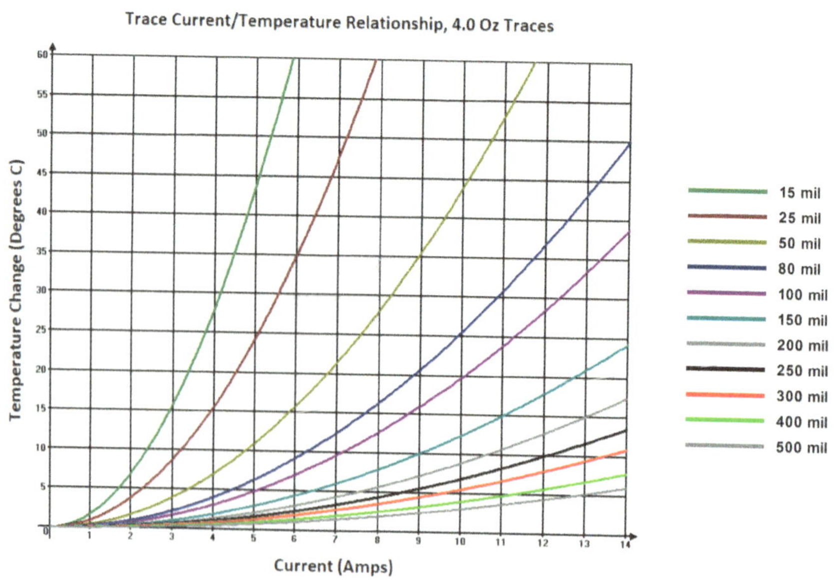

Figure E.6 4.0 Oz

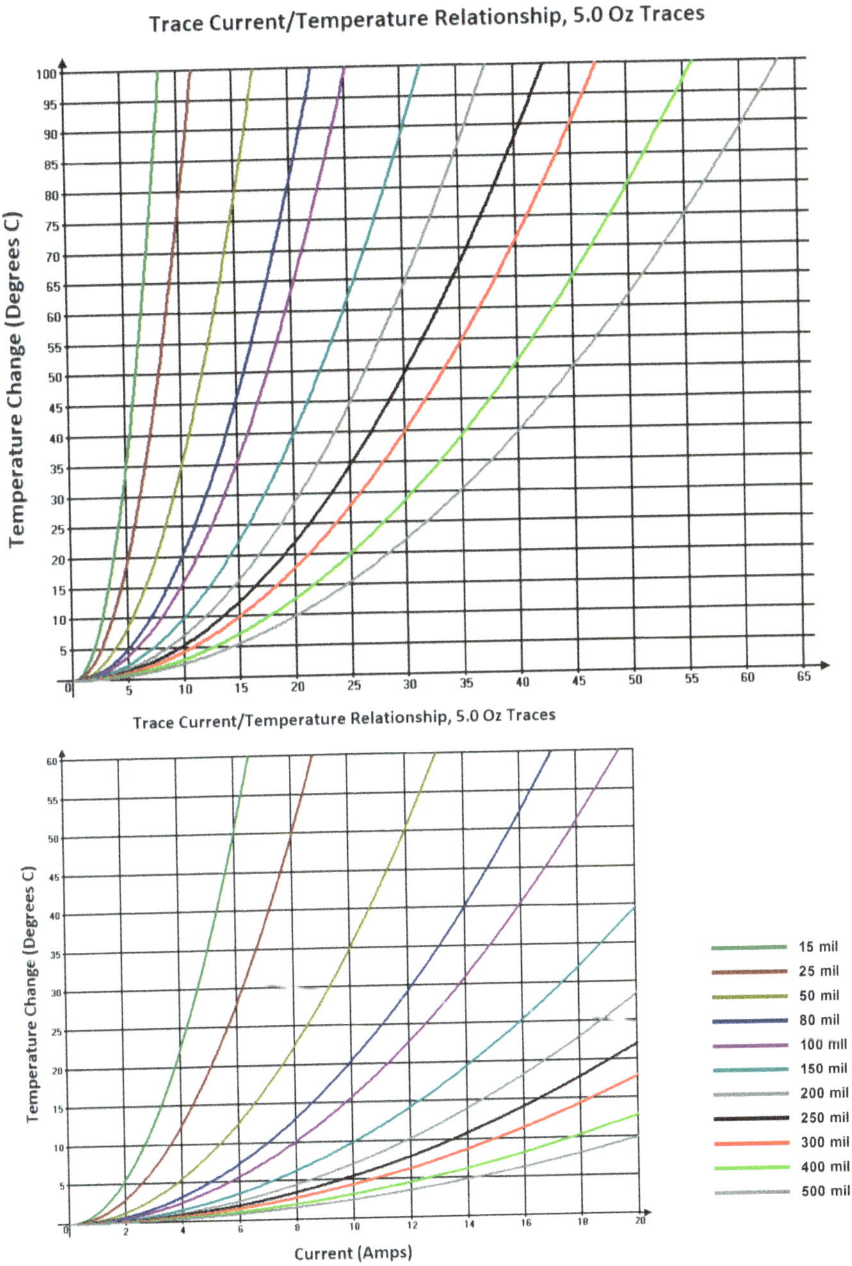

Figure E.7 5.0 Oz

APPENDIX

F

Contents

Current Density in Vias

F.1 Interpretations

The following sections look at the current densities (A/mm^2) at various layers in the via. Typically, the "Top" layer is the trace layer and Core 1 through Core 4 are the first four (of seven) dielectric layer models. Figure F.1 illustrates the analyses that are going on at each layer for each simulation. This figure is for the first core layer in the single via model.

At each layer, four current density readings are taken around each layer. They are labeled 1 through 4 in the column in the table. Position 1 corresponds to the left side of the via as shown in Figure F.1; 2 corresponds to the top of the via; 3 corresponds to the right-hand side of the via; and 4 corresponds to the bottom of the via. The density readings are recorded in the table and summed in the "Total" row. The average density ("Average" in the table) is the total of the density readings divided by 4 (the number of readings). The "x area" is the product of the average density times the via conducting area, which is always 0.0217 mm^2. This should equal the total current through the via (in this case 4.0 amps).

F.1.1 Caution

A major word of caution is required here. First, the model approximates the circular wall of the via with a series of square elements. We referred to these as thermal pixels (refer back to Section 6.3). Thus, the model approximates the cross-sectional conducting area of the via by adding up these thermal pixels. We calculate the conducting area as 0.0217 mm^2. The model's internal value might be slightly different. This is partly why the currents we calculate from the current densities might not exactly total the applied current.

Second, current density is a point concept. The density has a value at a specific point, and may vary from point-to-point. In order to obtain the total current in the via wall at any layer we must *integrate* the density around the circular wall of the via. What we are doing, taking only four readings, is an approximation. We believe four readings are reasonable because the variation in the density is smooth and continuous. But it is expected that our results will be approximate.

Third, note that in the via models, the current at the trace layer totals to more than the 4.0 amps applied to the trace. This effect diminishes as we look further down in the via. This is particularly apparent in the four-via models. This is because not all the current (density) measured at the front of the front vias actually enters the front (part of the) via. Some part of that current flows *past* the via and passes through the back side of the via or through a rear via. We can envision this more clearly if we think of the current as a series of packets (maybe electrons?). Some of the packets (electrons) flow right to the edge of the front via, but then flow past that

Core 1, 230 µm	
Position	Density
1	223.7
2	179.5
3	140.4
4	180.2
Total	723.8
Average	181.0
x area =	3.922 A

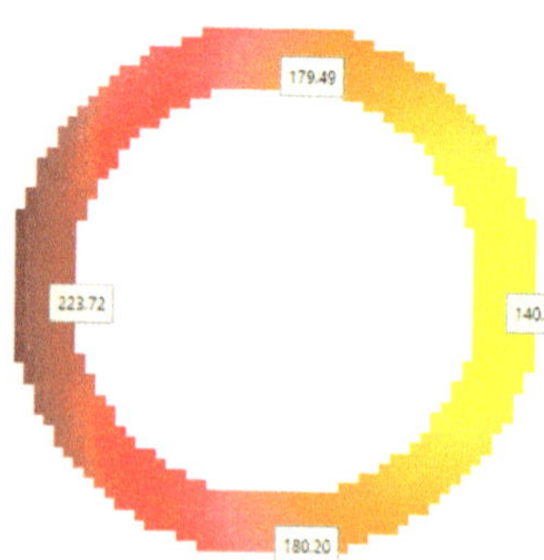

Figure F.1 Current densities around via wall, Core 1

point and enter the via at the back or enter a back via. Thus these packets (electrons) are counted twice in the current density measurements.

F.1.2 Symmetry

The current density patterns are symmetrical, so we need only look at the first four core layers. The remaining layers are identical in reverse order. In the case of four vias in the straight line configuration, the density patterns are symmetrical through Core 4 and also symmetrical around the centerline of the trace. In the case of four vias turning at an angle, there is a symmetry if the viewer looks at it as twisting as we go down through the via layers (what we call rotational symmetry). So in all cases we need only look at the core layers down through the middle one.

Figure F.2 illustrates the symmetry between the top (left) and bottom (right) layers in the single via model.

Figure F.3 illustrates what we mean by rotational symmetry. The horizontal trace (left) is the top trace layer and the vertical (right) trace is the bottom trace layer.

F.2 Single Via Model

Figures F4 through F8 illustrate the current density patterns for the single via model with traces going straight ahead.

F.3 Single Via Model with Core 1 Broken into Three Cores, the First Two with 15-μm Thicknesses

Note that the current density effects unique to the top (trace) layer (Figure F9) continue into the first small layer (Figure F10), and to a lesser extent into the second small layer (Figure F.11). After that, the current densities are essentially the same (Figures F.12 and F.13), as with the simulation shown in Section F.2.

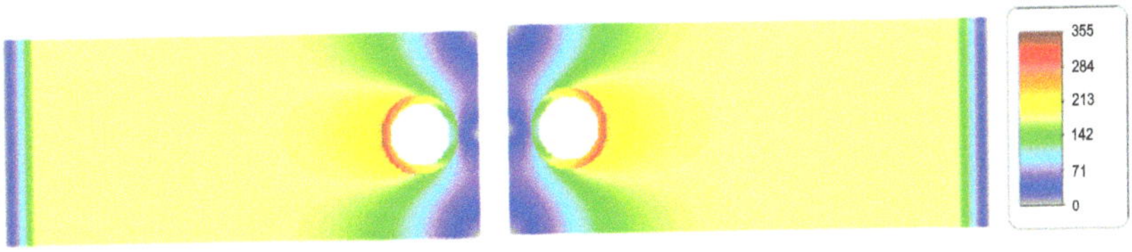

Figure F.2 Via current density symmetry between top (left) and bottom (right) layers.

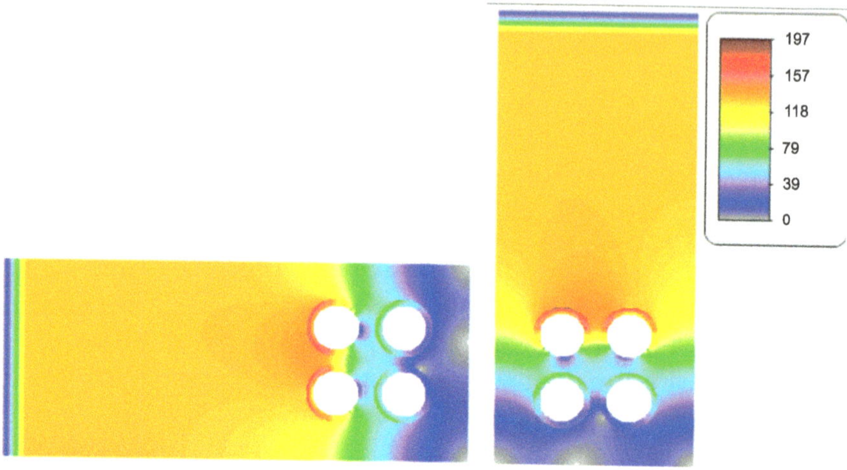

Figure F.3 Via current density symmetry for top (left) and bottom (right) layers for a corner via.

Position	Density
1	270.7
2	210.9
3	118.0
4	211.7
Total	811.3
Average	202.8
x area =	4.397 A

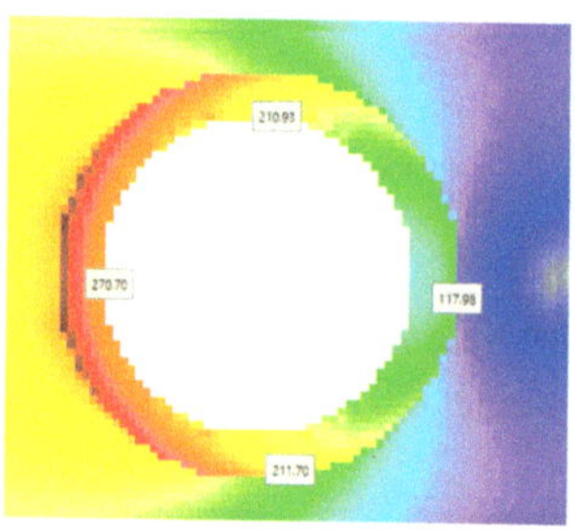

Figure F.4 Top layer, 34 μm thick.

Position	Density
1	223.7
2	179.5
3	140.4
4	180.2
Total	723.8
Average	181.0
x area =	3.922 A

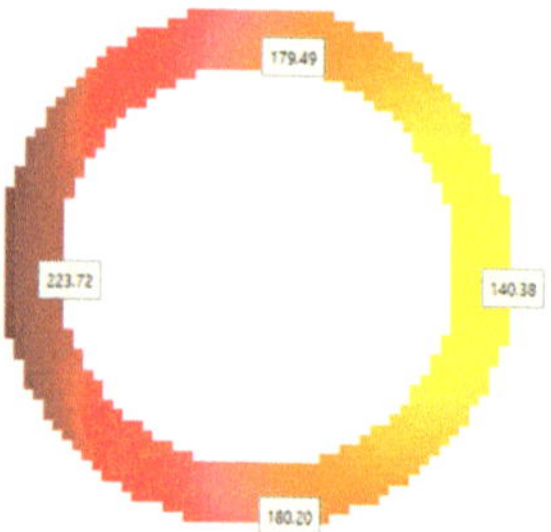

Figure F.5 Core 1, 230 μm thick.

Position	Density
1	189.2
2	181.2
3	173.8
4	181.6
Total	725.8
Average	181.5
x area =	3.933 A

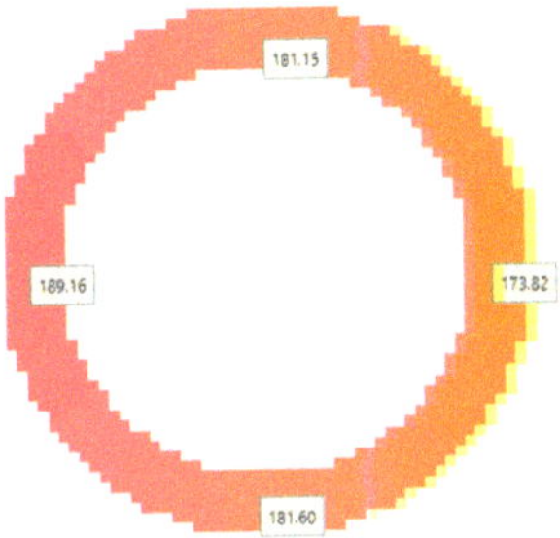

Figure F.6 Core 2, 230 μm thick.

Position	Density
1	182.7
2	181.3
3	180.0
4	181.4
Total	725.4
Average	181.4
x area =	3.931 A

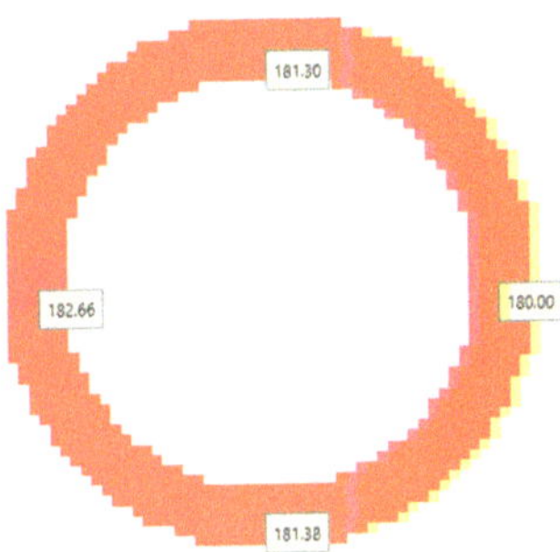

Figure F.7 Core 3, 230 μm thick.

Position	Density
1	181.3
2	181.3
3	181.3
4	181.3
Total	725.2
Average	181.3
x area =	3.93 A

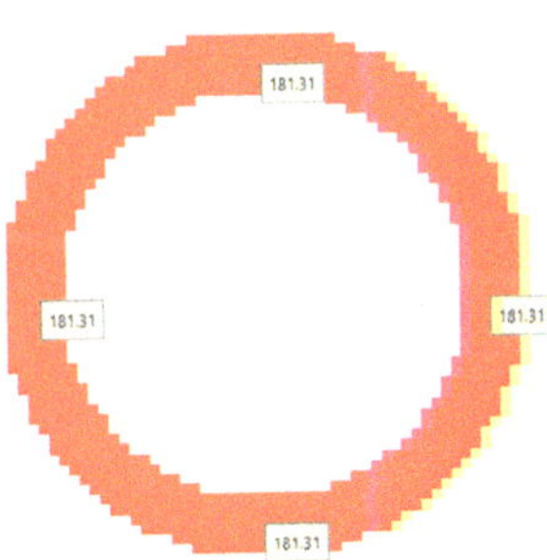

Figure F.8 Core 4, 230 μm thick.

Position	Density
1	288.2
2	186.4
3	108.6
4	200.9
Total	784.1
Average	196.03
x area =	4.249 A

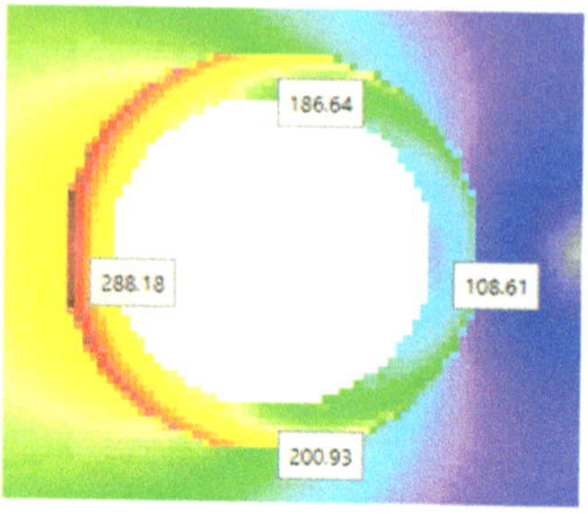

Figure F.9 Top, 34 μm thick.

Position	Density
1	256.8
2	184.5
3	107.0
4	191.5
Total	738.8
Average	184.7
x area =	4.004 A

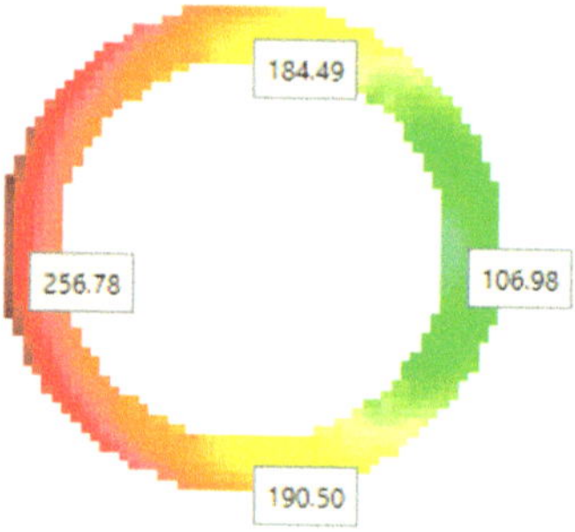

Figure F.10 Core 00, 15 μm thick.

Position	Density
1	248.0
2	190.9
3	121.0
4	191.9
Total	751.8
Average	187.95
x area =	4.074 A

Figure F.11 Core 01, 15 μm thick.

Position	Density
1	218.3
2	179.2
3	147.6
4	181.2
Total	726.3
Average	181.575
x area =	3.936 A

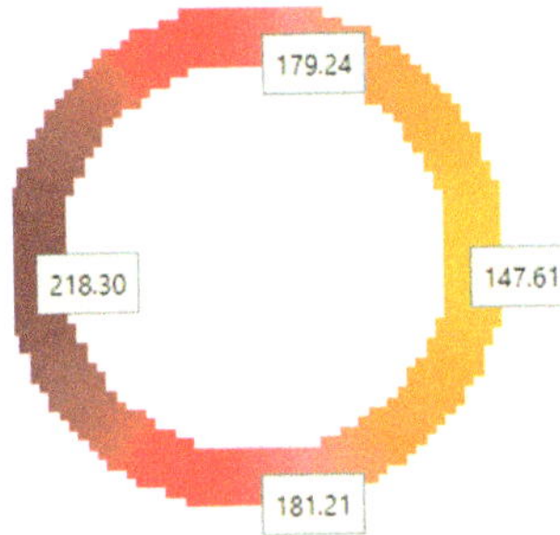

Figure F.12 Core 02, 200 μm thick.

Trace Layer (L)			Trace Layer (R)
Position	Density		Density
1	136.87		76.41
2	129.18		45.05
3	33.77		29.57
4	124.02		66.76
Total	423.84		217.90
Average	105.96		54.475
x area =	2.299 A		1.182 A

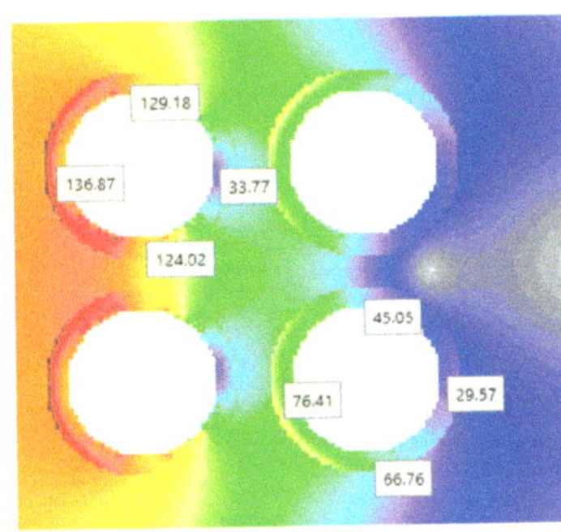

Figure F.13 Trace layer, 35 μm thick.

F.4 Simulation of Four Vias, Proceeding Straight Ahead

Figures F.13 through F.18 Illustrate the case of 4 vias in a traces going straight ahead. Note that the patterns are symmetrical around the horizontal center line of the trace.

F.5 Simulation of Four Vias, Traces at Right Angles

Figures F.19 through F.23 illustrate the current densities around the vias for the case of 4 vias with the connected traces at right-angles to each other.

Core 1 (L)			Core 1 (R)
Position	Density		Density
1	95.47		60.75
2	58.28		37.11
3	2.01		34.19
4	48.00		49.36
Total	203.76		181.51
Average	50.95		45.35
x area =	1.104 A		0.983 A

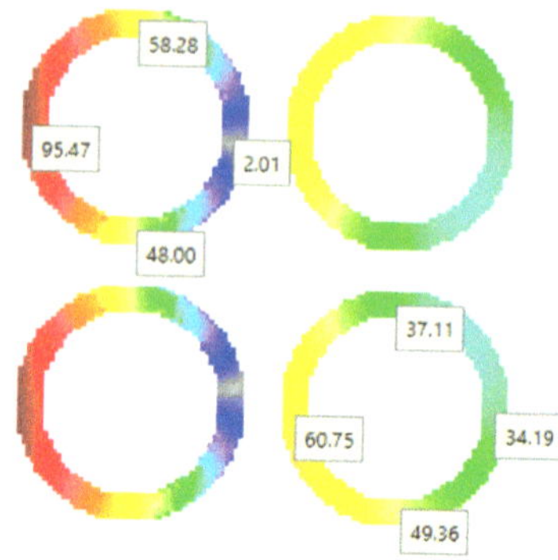

Figure F.14 Core 1, 230 μm thick.

Core 2 (L)			Core 2 (R)
Position	Density		Density
1	53.77		48.01
2	46.55		43.96
3	36.91		43.09
4	44.48		46.52
Total	171.71		181.58
Average	45.425		45.395
x area =	0.985 A		0.985 A

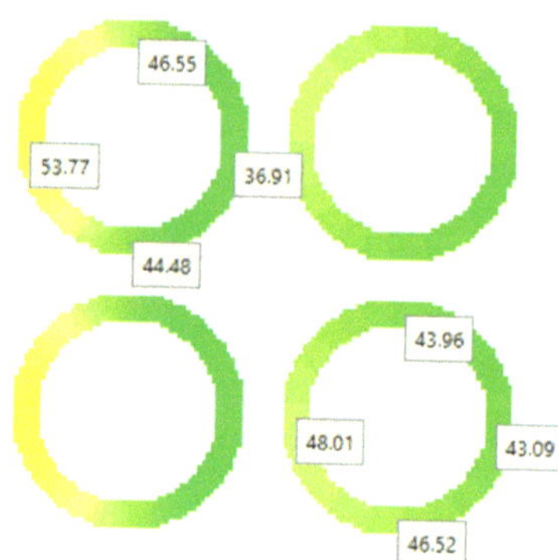

Figure F.15 Core 2, 230 μm thick.

Core 3 (L)			Core 3 (R)
Position	Density		Density
1	46.77		45.80
2	45.48		45.15
3	43.79		45.00
4	45.09		45.60
Total	181.13		181.55
Average	45.28		45.38
x area =	0.983 A		0.985 A

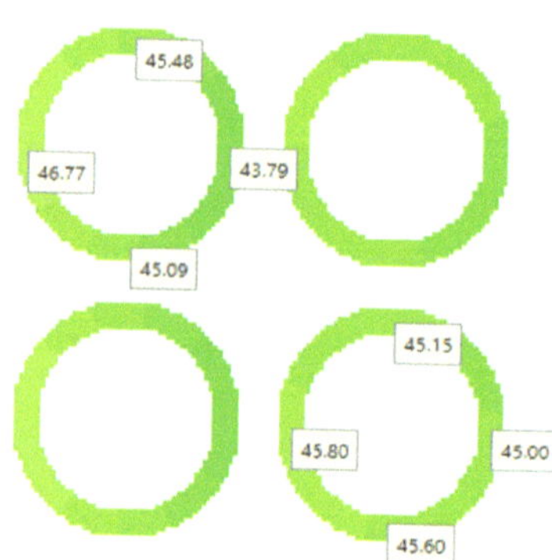

Figure F.16 Core 3, 230 μm thick.

Core 4 (L)			Core 4 (R)
Position	Density		Density
1	45.47		45.20
2	45.35		45.31
3	45.08		45.58
4	45.22		45.45
Total	181.12		181.54
Average	45.28		45.39
x area =	0.983 A		0.985 A

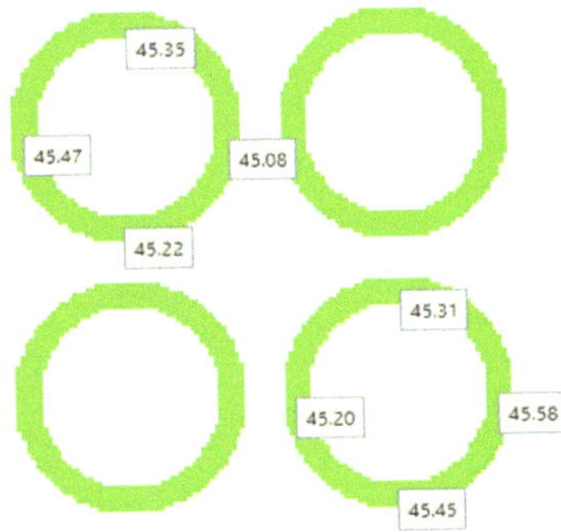

Figure F.17 Core 4, 230 μm thick.

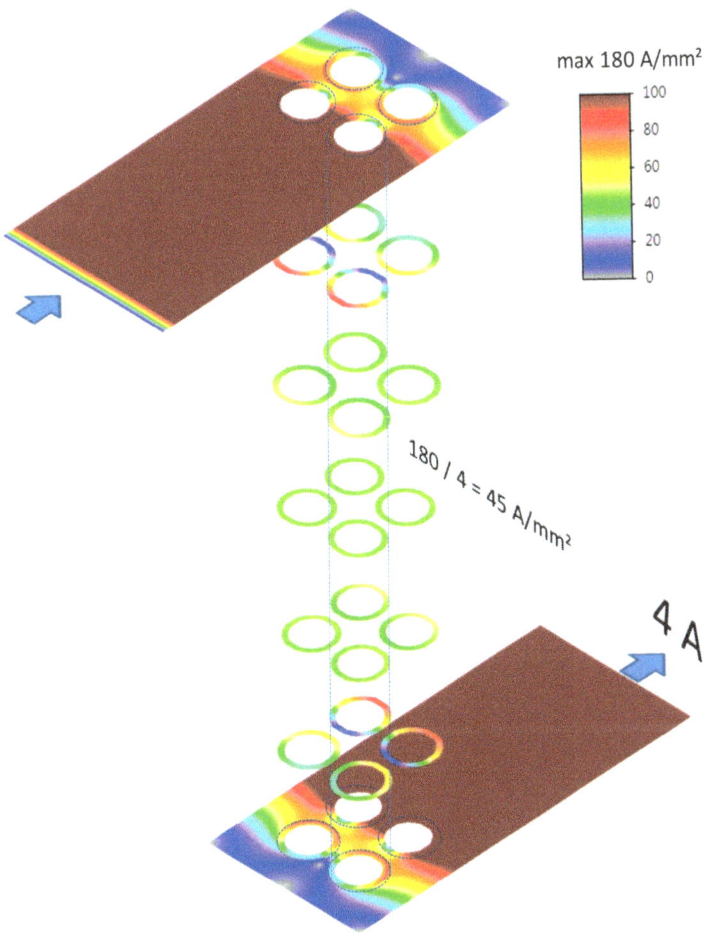

Figure F.18 3-D model of four-via simulation (courtesy of Dr. J. Adam.)

Top Layer, Upper Left		Top Layer, Upper Right
Position	Density	Density
1	144.01	71.14
2	127.53	62.43
3	21.84	28.13
4	124.12	43.09
Total	417.50	204.79
Average	104.38	51.20
x area =	2.263 A	1.10 A

Top Layer, Lower Left		Top Layer, Lower Right
Position	Density	Density
1	132.74	66.94
2	122.36	32.97
3	32.26	22.23
4	120.38	55.80
Total	407.74	177.94
Average	101.94	44.49
x area =	2.210 A	0.964 A

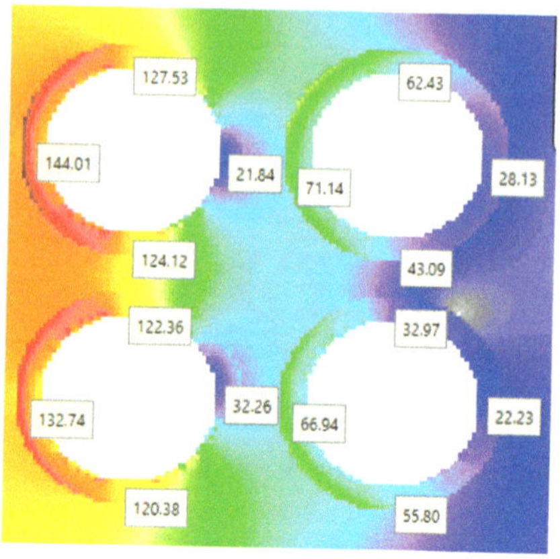

Figure F.19 Top layer, 35 μm thick.

Core 1, Upper Left		Core 1, Upper Right
Position	Density	Density
1	102.09	55.35
2	66.54	47.53
3	14.10	34.00
4	55.71	38.38
Total	238.44	175.26
Average	59.61	43.81
x area =	1.292 A	0.950 A

Core 1, Lower Left		Core 1, Lower Right
Position	Density	Density
1	89.62	47.05
2	45.43	24.52
3	7.47	24.06
4	57.65	39.63
Total	200.17	135.26
Average	50.05	33.82
x area =	1.085 A	0.733 A

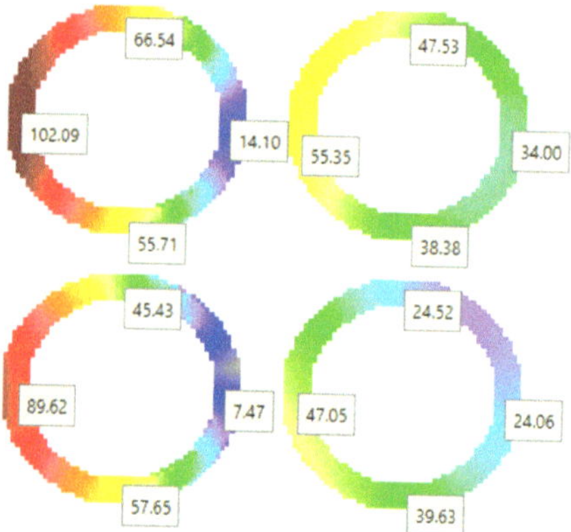

Figure F.20 Core 1, 230 μm thick.

Core 2, Upper Left			Core 2, Upper Right
Position	Density		Density
1	63.35		46.06
2	56.03		44.78
3	47.10		42.01
4	54.43		43.17
Total	220.91		176.02
Average	55.23		44.01
x area =	1.197 A		0.954 A

Core 2, Lower Left			Core 2, Lower Right
Position	Density		Density
1	52.51		35.80
2	43.09		31.86
3	36.66		31.30
4	45.92		34.76
Total	178.18		133.72
Average	44.55		33.43
x area =	0.966 A		0.725 A

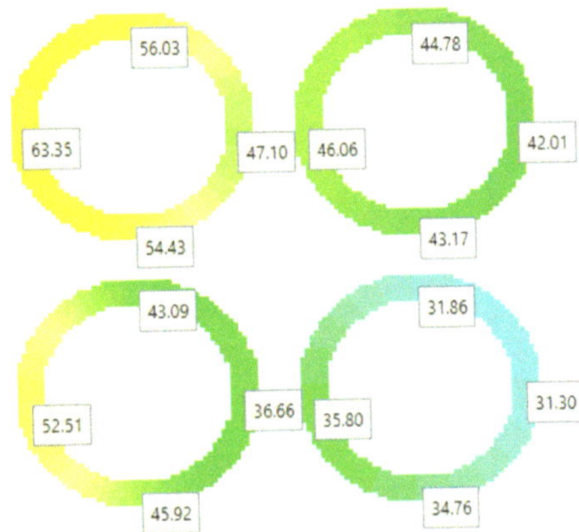

Figure F.21 Core 2, 230 μm thick.

Core 3, Upper Left			Core 3, Upper Right
Position	Density		Density
1	56.40		44.25
2	55.17		44.08
3	53.47		43.54
4	54.79		43.71
Total	219.83		175.58
Average	54.96		43.90
x area =	1.191 A		0.952 A

Core 3, Lower Left			Core 3, Lower Right
Position	Density		Density
1	45.57		33.65
2	43.83		32.99
3	42.73		32.86
4	44.32		33.49
Total	176.45		132.99
Average	44.11		33.25
x area =	0.956 A		0.721 A

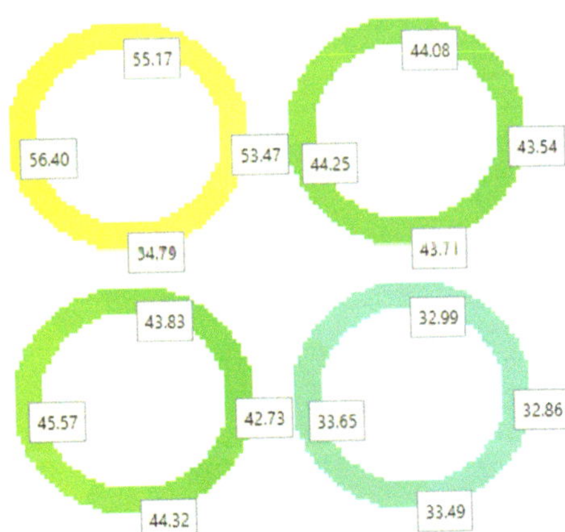

Figure F.22 Core 3, 230 μm thick.

Core 4, Upper Left		Core 4, Upper Right
Position	Density	Density
1	55.14	43.90
2	55.15	44.20
3	54.57	43.89
4	54.58	43.64
Total	219.44	175.63
Average	54.86	43.91
x area =	1.089 A	0.952 A

Core 4, Lower Left		Core 4, Lower Right
Position	Density	Density
1	44.29	33.22
2	44.03	33.22
3	43.73	33.18
4	43.99	33.17
Total	176.04	132.79
Average	44.01	33.20
x area =	0.954 A	0.720 A

Figure F.23 Core 4, 230 μm thick.

APPENDIX

G

Contents

Derivation of Onderdonk's Equation

G.1 Onderdonk's Equation

What follows is modified slightly from Johannes Adams' original derivation.[1] This derivation is complex and involves calculus. It is provided here:

1. So that some senior engineers who are interested may refer to it;
2. Because we are not aware of any other publications where it exists.

G.2 Background

We are aware of only one serious attempt at independently deriving Onderdonk's equation (there might well have been other attempts we are unaware of). Babrauskas and Wichman[2] did so with interesting results. Their Equation 6 derived the time to heat a copper wire to the melting temperature. Their Equation 7 derived the time to melt the wire. Their Equation 8 was the sum, which they then compared to Onderdonk's equation and remarked that their result was 17% higher than Onderdonk's. We believe that that comparison is not appropriate and they should have compered their Equation 6 with Onderdonk's. Our comparison of their Equation 6 with Onderdonk's suggests that their equation is just 4% below Onderdonk's, acceptably close.

G.2.1 Basic Equation

Deriving Onderdonk's equation is a complex matter. We start by equating two things. The first is the joule heating of a conductor by the electrical current, I^2R*time. The other is the gain of thermal energy of the conductor, mass * time * the specific heat capacity of the material. The equations then get more confused by the fact that the resistivity (of a material) and the density of a material are both represented by the same Greek symbol, ρ. So we differentiate those two symbols by letting ρ_{20} be the resistivity of copper (at 20°C) and ρ_d be the density of copper.

With those explanations in mind, then, Table G.1 is a list of variables used in the derivation.

In the following text, lhs means left-hand-side of the equation and rhs means right-hand-side of the equation. Equation G.1 describes an equality in a piece of matter over a time span Δt (starting at $t = 0$).

Gain of thermal energy (lhs) = joule heating by electric energy (rhs):

$$c_p \cdot M \cdot \Delta T = R \cdot I^2 \cdot \Delta t \tag{G.1}$$

Table G.1

Variables Used in Deriving Onderdonk's Equation

Symbol	Variable	Unit	Value
I	(DC) current	Ampere	
t	Time	Second	
T	Wire temperature (avg.)	°C	
Θ	Temperature rise $T–T_{ref}$	°C	
L	Length of wire	meter	
M	Mass of wire	kg	
R	Electric resistance of the wire	ohm	
c_p	Specific heat capacity (copper)	J/(kg*K)	385
ρ_d	Density of wire material (copper)	kg/m³	8900
ρ_{20}	Electric resistivity @20°C (copper)	ohm*m²/m	$1.72 * 10^{-8}$
α	Temperature coefficient of resistivity	1/°C	.00393@20°
A	Cross-sectional area of wire	m²	
K	Temporary utility variable		
δ	Another utility variable		

Because there is no heat loss over the surface, temperature on each side of this equation would rise infinitely with time. The units on both sides are $W \cdot s =$ Joule. M and R depend on the geometry of the wire, c_p is a material value.

The electric resistance of a wire is temperature-dependent. If the ambient temperature is 20°C, the resistance at any other temperature, T, is typically expressed by (G.2) (refer back to Chapter 3):

$$R = \rho_{20}(1 + \alpha_{20} \cdot (T - 20))\frac{L}{A} \tag{G.2}$$

Using the basic relation for volume and mass ($M = \rho_d{}^*L^*A$) and (G.2), we rewrite (G.1) as

$$c_p \cdot \rho_d \cdot L \cdot A \cdot \Delta T = \rho_{20}(1 + \alpha_{20}(T - 20))\frac{L}{A} \cdot I^2 \cdot \Delta t \tag{G.3}$$

The length of the wire on each side, L, will cancel. Also, we will divide through by A. If we assume the initial temperature of the wire as also being 20°C (see again note 3), we can unify ΔT and T-20 to a single variable temperature rise $\Theta = T - 20$. Because of the permanent link between t and Θ, $\Theta(t)$ will be a time-dependent variable and the correct basic equation is an ordinary differential equation:

$$c_p \cdot \rho_d \cdot \frac{d\Theta}{dt} = \rho_{20}(1 + \alpha_{20}\Theta(t))\frac{I^2}{A^2} \tag{G.4}$$

Note that I/A is current density (A/m^2).

G.2.2 Solving the Equation

We solve (G.4) step by step. First, we divide each side by $c_d \rho_d$. Then we collect some constant parameters in an auxiliary constant quantity K:

$$K = \frac{\rho_{20}}{c_p \cdot \rho_d}\frac{I^2}{A^2} \tag{G.5}$$

This, then, leads to

$$\frac{d\Theta}{dt} = K(1+\alpha_{20}\Theta) \tag{G.6}$$

Now if we define

$$\delta = 1 + \alpha_{20}\Theta \tag{G.7}$$

we get

$$\frac{d\Theta}{dt} = K\delta \tag{G.8}$$

This is an ordinary simple differential equation of the form

$$\frac{d\Theta}{dt} = \frac{d\Theta}{d\delta}\frac{d\delta}{dt} = K\delta \tag{G.9}$$

Recognizing (from (G.7)) that $\frac{d\Theta}{d\delta} = \frac{1}{\alpha_{20}}$

We finally get a simple ordinary differential equation:

$$\frac{d\delta}{dt} = \alpha_{20}K\delta \tag{G.10}$$

We set an initial condition that at time $t = 0$ the wire will be at 20°C. Therefore $\Theta(t = 0) = 0$, which leads to $\delta(t = 0) = 1$.

The way to solve (G.10) is to integrate[4]

$$\int_1^{\vartheta}\frac{d\delta}{\delta} = \int_0^t \alpha_{20}K\,dt \tag{G.11}$$

which gives

$$\ln(\delta) - \ln(1) = \alpha_{20}K(t-0) \tag{G.12}$$

or (because $ln(1) = 0$)

$$\ln(\delta) = \alpha_{20}Kt \tag{G.13}$$

We then unfold the auxiliary variables

$$\ln(1+\alpha_{20}\Theta) = \frac{\alpha_{20}\rho_{20}}{c_p \cdot \rho} \frac{I^2}{A^2} \cdot t \tag{G.14}$$

and exchange rhs and lhs (and move α to the denominator, for reasons that will be apparent in a moment):

$$\frac{\alpha_{20}\rho_{20}}{c_p \cdot \rho} \frac{I^2}{A^2} \cdot t = \ln\left(1+\frac{\Theta}{1/\alpha_{20}}\right) \tag{G.15}$$

We are going to solve the lhs of this equation separately. First, looking at

$$\frac{\alpha_{20}\rho_{20}}{c_p \rho_d} \frac{I^2}{A^2} \tag{G.16}$$

we rely on the relationship that $\alpha_{T2} * \rho_{T2} = \rho_{T1} * \alpha_{T1}$ at any two reference temperatures (for a given material). (See proof in Section G.3.) Therefore, we can use the 20 °C reference values for ρ and α regardless of the reference temperature for the equation. Note also that typical expressions for Onderdonk's equation (refer back to (11.3) and (11.4)) are in units of circular mils. So, our material data of copper are

$\rho_{20} = 0.0172\ \Omega\ \text{mm}^2/\text{m}$ (at 20°C) or $1.75 \cdot 10^{-8}\ \Omega\ \text{m}^2/\text{m}$

$\alpha_{20} = 0.00393$ 1/K, (at 20°C)

$c_p = 385$ J/kgK

$\rho_d = 8900\ \text{kg/m}^3$

To calculate the coefficient in (G.15) in front of I^2/A^2 let us use our material data and convert (G.15) to circular-mils:

a. First : $\frac{\alpha_{20} \cdot \rho_{20}}{c_p \cdot \rho_d}$ = (0.00393 * 1.72 * 10^{-8})/(385 * 8900) = 1.973 * 10^{-17}

b. Then: since A is in square meters, and 1.0 m^2 = 1.98 * 10^9 circ-mils, therefore, A^2 = 3.9 * 10^{18} circ-mils

c. So introducing (b) into (a) → $1.973 \cdot 10^{-17}$ * 3.9 * 10^{18} = 76.9

This all leads to

$$76.9 * \frac{I^2}{A^2} * t = \ln\left(1 + \frac{\Theta}{1/\alpha_{20}}\right) \tag{G.17}$$

To convert from the natural log, ln, to $\log_{10}$, we have to divide by ln(10) = 2.30. This leads to 76.9/2.30 = 33.5.

So, with *A* in circular mils, (G.17) becomes

$$33.5\left(\frac{I}{A}\right)^2 t = \log_{10}\left(1 + \frac{\Theta}{1/\alpha_{20}}\right) \tag{G.18}$$

Now, to solve for $1/\alpha$ for any reference temperature, we note again that $\alpha_{T2} * \rho_{T2} = \rho_{T1} * \alpha_{T1}$

This leads to

$$1/\alpha_{T2} = \rho_{T2}/(\rho_{T1}\alpha_{T1}) = (\rho_{T1} + \rho_{T1}\alpha_{T1}(T2 - T1))/\rho_{T1}\alpha_{T1} \tag{G.19}$$

or

$$1/\alpha_{T2} = \frac{\rho_{T1}}{\rho_{T1}\alpha_{T1}} + (T2 - T1) = \frac{1}{\alpha_{T1}} + (T2 - T1) \tag{G.20}$$

If we let *T*1 = ambient = 20°C and *T*2 be any other reference temperature, T_{ref}, then

$$1/\alpha_{ref} = \frac{1}{\alpha_{20}} + T_{ref} - 20 = 254 - 20 + T_{ref} = 234 + T_{ref} \tag{G.21}$$

Leading to our final expression

$$33.5\left(\frac{I}{A}\right)^2 t = \log_{10}\left(\frac{\Theta}{234 + T_{ref}} + 1\right) \tag{G.22}$$

which at a reference temperature of 40°C (Onderdonk's reference) is

$$33.5\left(\frac{I}{A}\right)^2 t = \log_{10}\left(\frac{\Theta}{274} + 1\right) \tag{G.23}$$

or equal to (11.3) and (11.4).

It is sometimes desirable to have(G.23) expressed in terms of t, the time to reach the melting temperature. Solving (G.22) for t leads to

$$t = \left(\frac{1}{33.5}\right)\left[\log_{10}\left(\frac{\Theta}{234 + T_{ref}} + 1\right) * \left(\frac{A}{I}\right)^2\right] \qquad (G.24)$$

where A is still in circular mils.

Using(G.24), we can now transform Onderdonk's equation to any temperature reference point. We offer it in Table G2 with reference temperatures of 20°C, 40°C, 85°C, and 105°C without further calculations. The latter temperature references are important in the automotive industries. (Note that the higher the reference temperature the shorter the time is to reach the melting temperature.)

At the melting point of copper (1083°C) and at an initial temperature of 20°C,(G.24) can be reduced to the form

$$t = c * (A/I)^2 \qquad (G.25)$$

where $c = .0213$ in circular mils.

The values for c are supplied in Table G.3 for various units and reference temperatures.

G.3 Proof that $\alpha_{T2} * \rho_{T2} = \rho_{T1} * \alpha_{T1}$

Start with the standard relationship for converting resistivity from one reference temperature to another (see Chapter 3):

Table G.2
Log_{10} Denominator as a Function of Reference Temperature

Reference Temp (°C)	Denominator in Log Expression
20	254
40	274
85	319
105	339

Table G.3
Value of *c* for Various Reference Temperatures and Units

Reference Temperature	*c* in Circular mils	*c* in Square mils	*c* in Square mm
20	.0213	.0346	8.30×10^4
40	.0203	.0330	7.92
85	.0184	.0298	7.15
105	.0176	.0285	6.85

$$\frac{\rho_T}{\rho_{Tref}} = 1 + \alpha_{Tref}\left(T - T_{ref}\right) \qquad \text{(G.26)}$$

$$\rho_T = \rho_{Tref} + \rho_{Tref} * \alpha_{Tref}\left(T - T_{ref}\right) \qquad \text{(G.27)}$$

Let *T*1, *T*2, and *T*3 be any three temperatures. Then the conversion of resistivity between any two temperatures is calculated as follows:

a. T1 to T2: $\rho_{T2} = \rho_{T1} + \rho_{T1} * \alpha_{T1}\left(T2 - T1\right)$ (G.28)

b. T1 to T3: $\rho_{T3} = \rho_{T1} + \rho_{T1} * \alpha_{T1}\left(T3 - T1\right)$ (G.29)

c. T2 to T3: $\rho_{T3} = \rho_{T2} + \rho_{T2} * \alpha_{T2}\left(T3 - T2\right)$ (G.30)

Note that (G.29) must equal (G.30). We can expand(G.29) as follows:

$$\rho_{T3} = \rho_{T1} + \rho_{T1} * \alpha_{T1}\left(T3 - T2\right) + \rho_{T1} * \alpha_{T1}\left(T2 - T1\right) \qquad \text{(G.31)}$$

Now we set this equation equal to(G.30):

$$\begin{aligned} &\rho_{T1} + \rho_{T1} * \alpha_{T1}\left(T3 - T2\right) + \rho_{T1} * \alpha_{T1}\left(T2 - T1\right) = \\ &\rho_{T2} + \rho_{T2} * \alpha_{T2}\left(T3 - T2\right) \end{aligned} \qquad \text{(G.32)}$$

Substitute ρ_{T2} in (G.32):

$$\begin{aligned} &\rho_{T2} + \rho_{T1} * \alpha_{T1}\left(T3 - T2\right) = \\ &\rho_{T2} + \rho_{T2} * \alpha_{T2}\left(T3 - T2\right) \end{aligned} \qquad \text{(G.33)}$$

Subtract ρ_{T2} and divide each side by ($T3$–$T2$) leaves

$$\rho_{T1} * \alpha_{T1} = \rho_{T2} * \alpha_{T2} \tag{G.34}$$

or any two reference temperatures.

End Notes

1. Adam, J., white paper No.10, www.adam-research.de/pdfs/TRM_WhitePaper10_AdiabaticWire.pdf.
2. Babrauskas, V., and I. Wichman, "Fusing of Wires by Electrical Current," *Fire and Materials 2011 Conference Proceedings*, Interscience Communications (download a copy from https://www.academia.edu/9357019/Fusing_of_Wires_by_Electrical_Current_or from www.ultracad.com/articles/reprints/babrauskas.pdf.
3. From a thermodynamic engineering standpoint, there is a significant difference between the ambient temperature and the initial condition. For example, there could be some heating of a wire above the ambient temperature in a normal operation before a change of current is applied. In this book we are assuming that the initial condition (temperature) is the same as the ambient temperature.
4. For those of us who have forgotten: $\int \frac{dx}{x} = Ln(x)$

APPENDIX H

Results of All Six Fusing Configuration Simulations

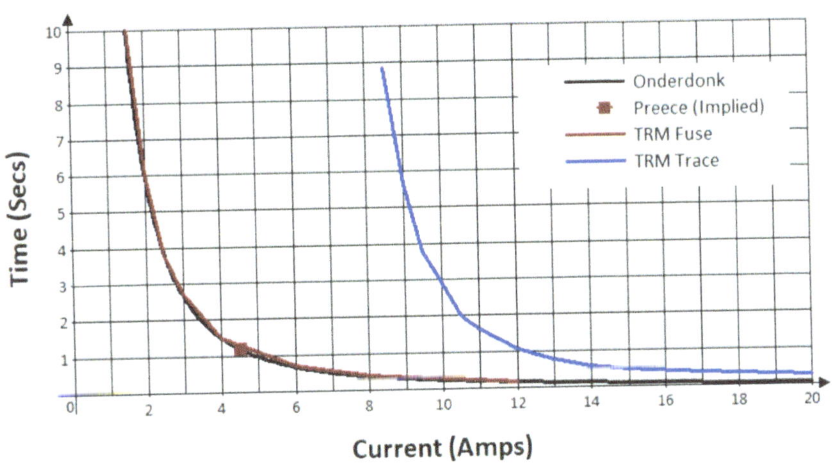

Figure H.1 1.0-oz, 20-mil fusing simulations.

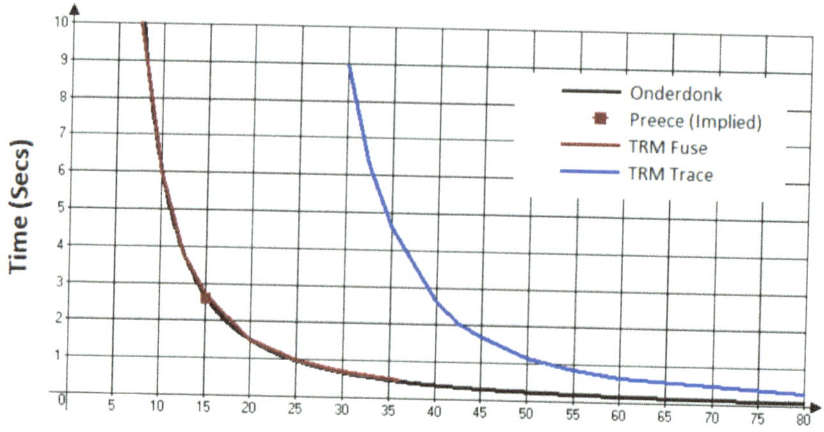

Figure H.2 1.0-oz, 100-mil fusing simulations.

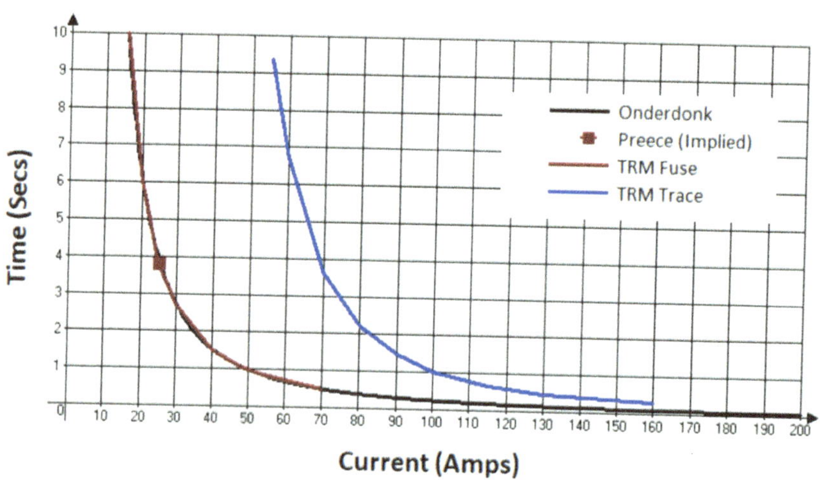

Figure H.3 1.0-oz, 200-mil fusing simulations.

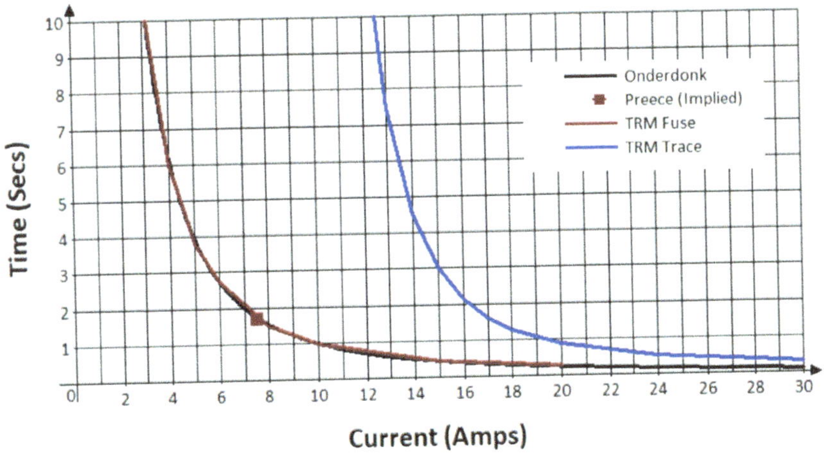

Figure H.4 2.0-oz, 20-mil fusing simulations.

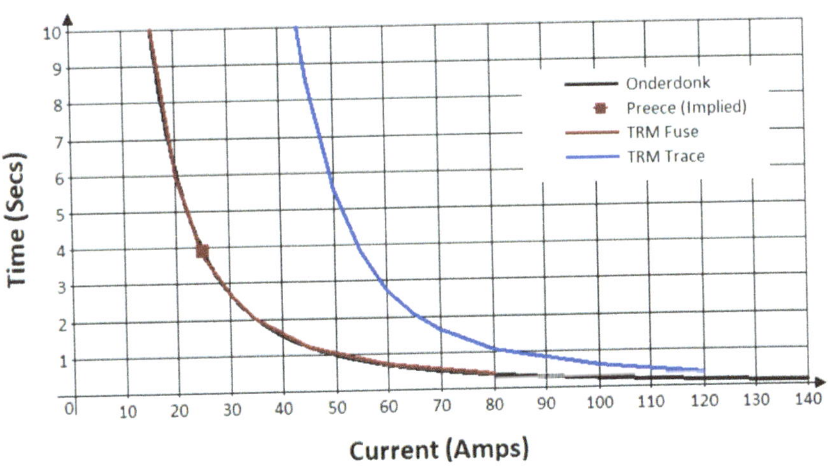

Figure H.5 2.0-oz, 100-mil fusing simulations.

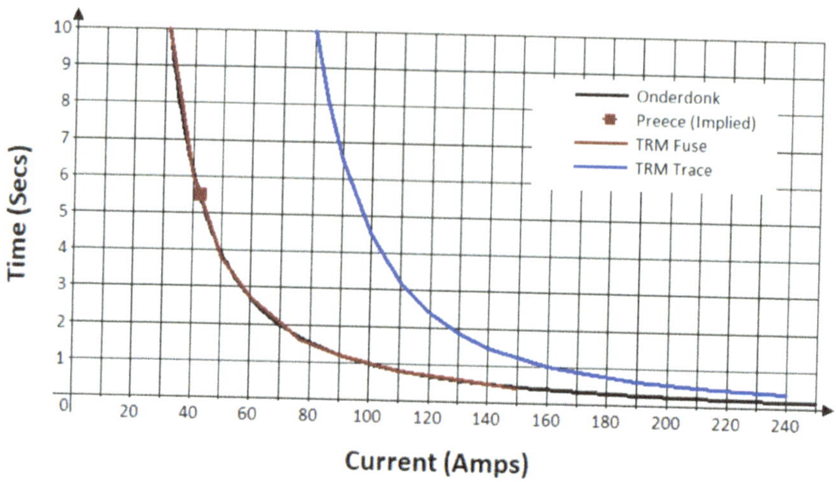

Figure H.6 2.0-oz, 200-mil fusing simulations.

APPENDIX

I

Nonuniform Heating Patterns

The following images are close-ups of the thermal distribution on the selected traces summarized in Table 13.1. The traces carried enough current to heat the trace to approximately 55°C, except where the trace was large enough that 10 amps could not reach that temperature (10 amps was the maximum output of the available current generator). The images cover a very narrow temperature range that is indicated by the scale under the image.

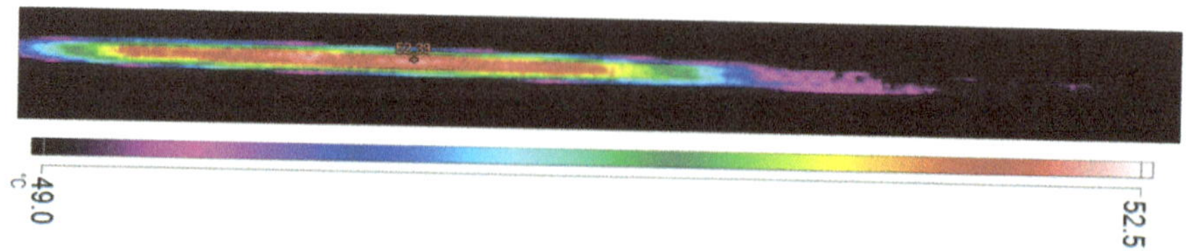

Figure I.1 100-mil wide, 0.5-oz foil trace.

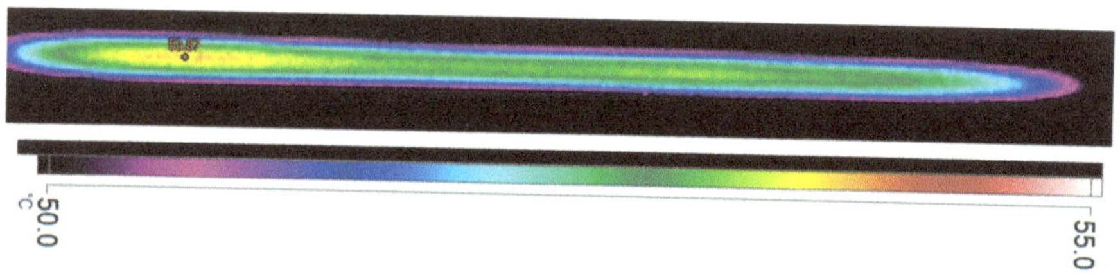

Figure I.2 100 mil wide 0.5 Oz. foil trace plated with 1.0 Oz. coper.

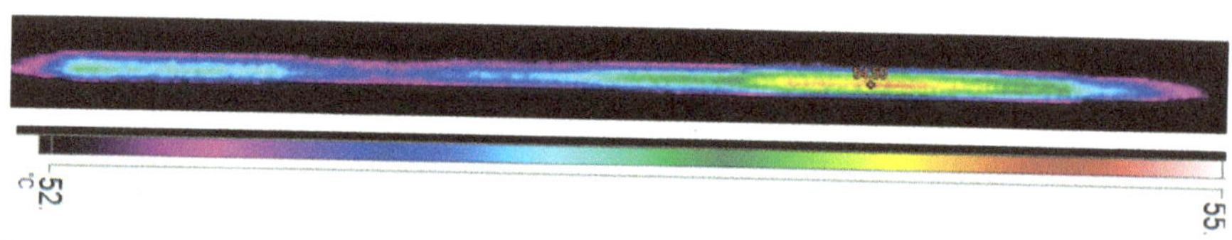

Figure I.3 100 mil wide 1.5 Oz. foil trace.

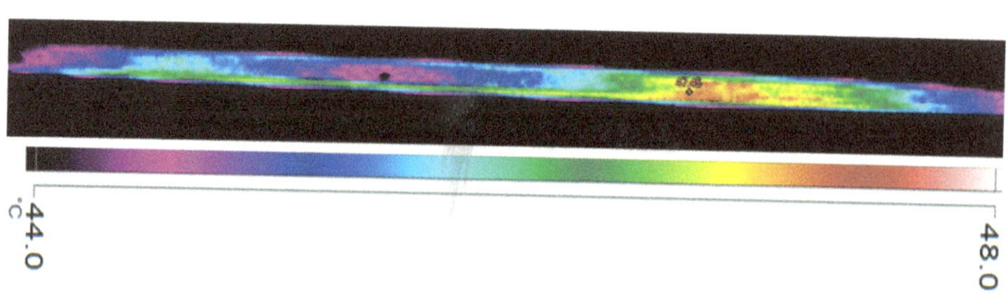

Figure I.4 100 mil wide 1.5 Oz. foil trace plated with 1.75 Oz. coper

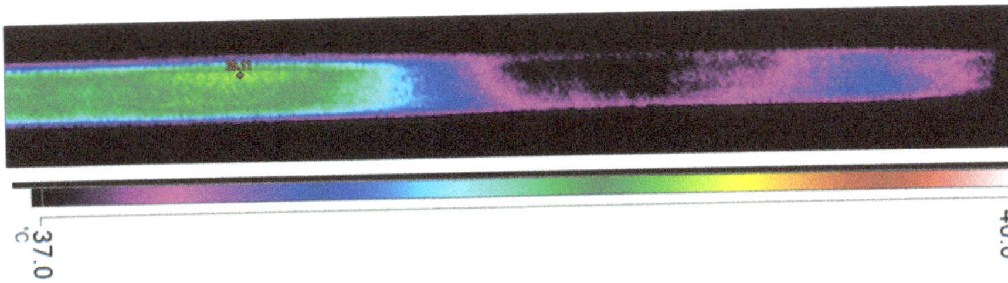

Figure I.5 200 mil wide 2.0 Oz. foil trace.

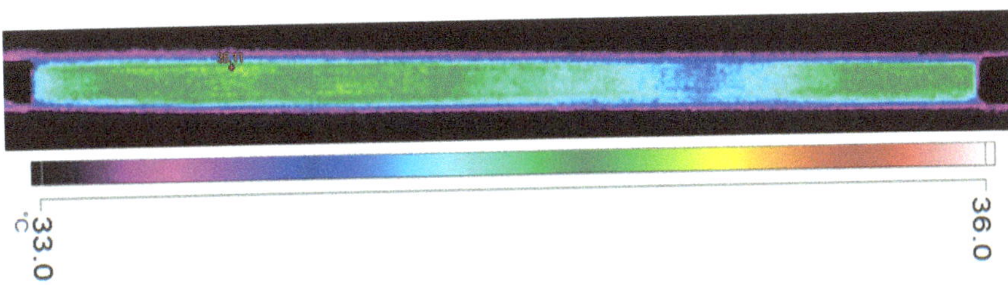

Figure U.6 200 mil wide 2.0 Oz. foil trace plated with 1.0 Oz. coper.

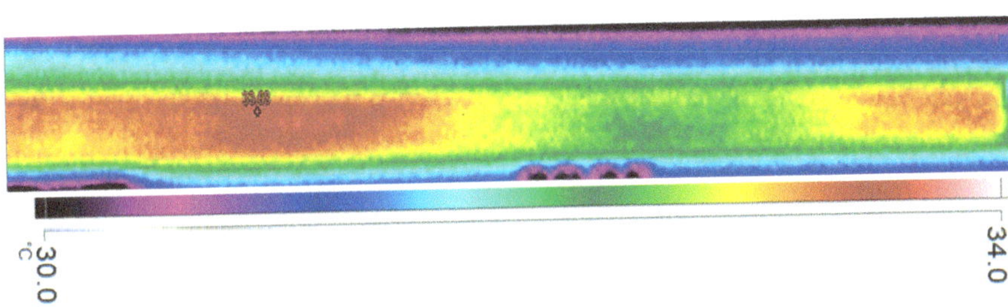

Figure I.7 200 mil wide 2.0 Oz. foil trace plated with 2.0 Oz. coper.

About the Authors

Douglas Brooks has a B.S./E.E. and an M.S./E.E. from Stanford and a Ph.D. from the University of Washington. In his final 20 years in business he owned a small engineering service firm and wrote numerous technical articles on printed circuit board design and signal integrity issues. He has published two books on these topics. He has given seminars several times a year all over the United States, as well as Moscow, China, Taiwan, Japan, Israel, and Canada. His primary focus is on making complex technical issues easily understood by those without advanced degrees. He is currently retired and lives with his wife of 53 years near Seattle, Washington.

Johannes Adam received a doctorate in physics from University of Heidelberg, Germany, in 1989 on a thesis about numerical treatment of three-dimensional radiation transport in moving astrophysical plasmas. He was then employed in software companies, mainly working on numerical simulations of electronics cooling at companies like Cisi Ingenierie S.A., Flomerics. Ltd., and Mentor Graphics Corp. In 2009 he founded ADAM Research and does work as a technical consultant for electronics developing companies and as a software developer. He is the author of a simulation program called Thermal Risk Management (TRM), designed for electronics developers and PCB designers who want to solve electrothermal problems at the board level. He is a member of the German chapter of IPC (FED e.V.) and engages in its seminars about thermal topics. He is a certified interconnect designer (CID). Dr. Adam lives in Leimen, near Heidelberg.

Index

N

O

P

R

S

T

Artech House Microwave Library

Behavioral Modeling and Linearization of RF Power Amplifiers, John Wood

Chipless RFID Reader Architecture, Nemai Chandra Karmakar, Prasanna Kalansuriya, Randika Koswatta, and Rubayet E-Azim

Control Components Using Si, GaAs, and GaN Technologies, Inder J. Bahl

Design of Linear RF Outphasing Power Amplifiers, Xuejun Zhang, Lawrence E. Larson, and Peter M. Asbeck

Design Methodology for RF CMOS Phase Locked Loops, Carlos Quemada, Guillermo Bistué, and Iñigo Adin

Design of CMOS Operational Amplifiers, Rasoul Dehghani

Design of RF and Microwave Amplifiers and Oscillators, Second Edition, Pieter L. D. Abrie

Digital Filter Design Solutions, Jolyon M. De Freitas

Discrete Oscillator Design Linear, Nonlinear, Transient, and Noise Domains, Randall W. Rhea

Distortion in RF Power Amplifiers, Joel Vuolevi and Timo Rahkonen

Distributed Power Amplifiers for RF and Microwave Communications, Narendra Kumar and Andrei Grebennikov

Electric Circuits: A Primer, J. C. Olivier

Electronics for Microwave Backhaul, Vittorio Camarchia, Roberto Quaglia, and Marco Pirola, editors

EMPLAN: Electromagnetic Analysis of Printed Structures in Planarly Layered Media, Software and User's Manual, Noyan Kinayman and M. I. Aksun

An Engineer's Guide to Automated Testing of High-Speed Interfaces, Second Edition, José Moreira and Hubert Werkmann

Envelope Tracking Power Amplifiers for Wireless Communications, Zhancang Wang

Essentials of RF and Microwave Grounding, Eric Holzman

Frequency Measurement Technology, Ignacio Llamas-Garro, Marcos Tavares de Melo, and Jung-Mu Kim

FAST: Fast Amplifier Synthesis Tool—Software and User's Guide, Dale D. Henkes

Feedforward Linear Power Amplifiers, Nick Pothecary

Filter Synthesis Using Genesys S/Filter, Randall W. Rhea

Foundations of Oscillator Circuit Design, Guillermo Gonzalez

Frequency Synthesizers: Concept to Product, Alexander Chenakin

Fundamentals of Nonlinear Behavioral Modeling for RF and Microwave Design, John Wood and David E. Root, editors

Generalized Filter Design by Computer Optimization, Djuradj Budimir

Handbook of Dielectric and Thermal Properties of Materials at Microwave Frequencies, Vyacheslav V. Komarov

Handbook of RF, Microwave, and Millimeter-Wave Components, Leonid A. Belov, Sergey M. Smolskiy, and Victor N. Kochemasov

High-Efficiency Load Modulation Power Amplifiers for Wireless Communications, Zhancang Wang

High-Linearity RF Amplifier Design, Peter B. Kenington

High-Speed Circuit Board Signal Integrity, Second Edition, Stephen C. Thierauf

Integrated Microwave Front-Ends with Avionics Applications, Leo G. Maloratsky

Intermodulation Distortion in Microwave and Wireless Circuits, José Carlos Pedro and Nuno Borges Carvalho

Introduction to Modeling HBTs, Matthias Rudolph

An Introduction to Packet Microwave Systems and Technologies, Paolo Volpato

Introduction to RF Design Using EM Simulators, Hiroaki Kogure, Yoshie Kogure, and James C. Rautio

Introduction to RF and Microwave Passive Components, Richard Wallace and Krister Andreasson

Klystrons, Traveling Wave Tubes, Magnetrons, Crossed-Field Amplifiers, and Gyrotrons, A. S. Gilmour, Jr.

Lumped Elements for RF and Microwave Circuits, Inder Bahl

Lumped Element Quadrature Hybrids, David Andrews

Microstrip Lines and Slotlines, Third Edition, Ramesh Garg, Inder Bahl, and Maurizio Bozzi

Microwave Circuit Modeling Using Electromagnetic Field Simulation, Daniel G. Swanson, Jr. and Wolfgang J. R. Hoefer

Microwave Component Mechanics, Harri Eskelinen and Pekka Eskelinen

Microwave Differential Circuit Design Using Mixed-Mode S-Parameters, William R. Eisenstadt, Robert Stengel, and Bruce M. Thompson

Microwave Engineers' Handbook, Two Volumes, Theodore Saad, editor

Microwave Filters, Impedance-Matching Networks, and Coupling Structures, George L. Matthaei, Leo Young, and E. M. T. Jones

Microwave Imaging Methods and Applications, Matteo Pastorino and Andrea Randazzo

Microwave Material Applications: Device Miniaturization and Integration, David B. Cruickshank

Microwave Materials and Fabrication Techniques, Second Edition, Thomas S. Laverghetta

Microwave Materials for Wireless Applications, David B. Cruickshank

Microwave Mixer Technology and Applications, Bert Henderson and Edmar Camargo

Microwave Mixers, Second Edition, Stephen A. Maas

Microwave Network Design Using the Scattering Matrix, Janusz A. Dobrowolski

Microwave Power Amplifier Design with MMIC Modules, Howard Hausman

Microwave Radio Transmission Design Guide, Second Edition, Trevor Manning

Microwave and RF Semiconductor Control Device Modeling, Robert H. Caverly

Microwave Transmission Line Circuits, William T. Joines, W. Devereux Palmer, and Jennifer T. Bernhard

Microwaves and Wireless Simplified, Third Edition, Thomas S. Laverghetta

Modern Microwave Circuits, Noyan Kinayman and M. I. Aksun

Modern Microwave Measurements and Techniques, Second Edition, Thomas S. Laverghetta

Modern RF and Microwave Filter Design, Protap Pramanick and Prakash Bhartia

Neural Networks for RF and Microwave Design, Q. J. Zhang and K. C. Gupta

Noise in Linear and Nonlinear Circuits, Stephen A. Maas

Nonlinear Microwave and RF Circuits, Second Edition, Stephen A. Maas

On-Wafer Microwave Measurements and De-Embedding, Errikos Lourandakis

Parameter Extraction and Complex Nonlinear Transistor Models, Günter Kompa

Passive RF Component Technology: Materials, Techniques, and Applications, Guoan Wang and Bo Pan, editors

PCB Design Guide to Via and Trace Currents and Temperatures, Douglas Brooks with Johannes Adam

Practical Analog and Digital Filter Design, Les Thede

Practical Microstrip Design and Applications, Günter Kompa

Practical Microwave Circuits, Stephen Maas

Practical RF Circuit Design for Modern Wireless Systems, Volume I: Passive Circuits and Systems, Les Besser and Rowan Gilmore

Practical RF Circuit Design for Modern Wireless Systems, Volume II: Active Circuits and Systems, Rowan Gilmore and Les Besser

Principles of RF and Microwave Design, Matthew A. Morgan

Production Testing of RF and System-on-a-Chip Devices for Wireless Communications, Keith B. Schaub and Joe Kelly

Q Factor Measurements Using MATLAB , Darko Kajfez

QMATCH: Lumped-Element Impedance Matching, Software and User's Guide, Pieter L. D. Abrie

Radio Frequency Integrated Circuit Design, Second Edition, John W. M. Rogers and Calvin Plett

Reflectionless Filters, Matthew A. Morgan

RF Bulk Acoustic Wave Filters for Communications, Ken-ya Hashimoto

RF Circuits and Applications for Practicing Engineers, Mouqun Dong

RF Design Guide: Systems, Circuits, and Equations, Peter Vizmuller

RF Linear Accelerators for Medical and Industrial Applications, Samy Hanna

RF Measurements of Die and Packages, Scott A. Wartenberg

The RF and Microwave Circuit Design Handbook, Stephen A. Maas

RF and Microwave Coupled-Line Circuits, Rajesh Mongia, Inder Bahl, and Prakash Bhartia

RF and Microwave Oscillator Design, Michal Odyniec, editor

RF Power Amplifiers for Wireless Communications, Second Edition, Steve C. Cripps

RF Systems, Components, and Circuits Handbook, Ferril A. Losee

Scattering Parameters in RF and Microwave Circuit Analysis and Design, Janusz A. Dobrowolski

The Six-Port Technique with Microwave and Wireless Applications, Fadhel M. Ghannouchi and Abbas Mohammadi

Solid-State Microwave High-Power Amplifiers, Franco Sechi and Marina Bujatti

Spin Transfer Torque Based Devices, Circuits, and Memory, Brajesh Kumar Kaushik and Shivam Verma

Stability Analysis of Nonlinear Microwave Circuits, Almudena Suárez and Raymond Quéré

Substrate Noise Coupling in Analog/RF Circuits, Stephane Bronckers, Geert Van der Plas, Gerd Vandersteen, and Yves Rolain

System-in-Package RF Design and Applications, Michael P. Gaynor

Technologies for RF Systems, Terry Edwards

Terahertz Metrology, Mira Naftaly, editor

TRAVIS 2.0: Transmission Line Visualization Software and User's Guide, Version 2.0, Robert G. Kaires and Barton T. Hickman

Understanding Microwave Heating Cavities, Tse V. Chow Ting Chan and Howard C. Reader

Understanding Quartz Crystals and Oscillators, Ramón M. Cerda

Vertical GaN and SiC Power Devices, Kazuhiro Mochizuki

The VNA Applications Handbook, Gregory Bonaguide and Neil Jarvis

Wired and Wireless Seamless Access Systems for Public Infrastructure, Tetsuya Kawanishi

For further information on these and other Artech House titles, including previously considered out-of-print books now available through our In-Print-Forever® (IPF®) program, contact:

Artech House Publishers
685 Canton Street
Norwood, MA 02062
Phone: 781-769-9750
Fax: 781-769-6334
e-mail: artech@artechhouse.com

Artech House Books
16 Sussex Street
London SW1V 4RW UK
Phone: +44 (0)20 7596 8750
Fax: +44 (0)20 7630 0166
e-mail: artech-uk@artechhouse.com

Find us on the World Wide Web at: www.artechhouse.com